BASICS OF MOLECULAR BIOLOGY

BASICS OF MOLECULAR BIOLOGY

SHILPA PANDEY

ANMOL PUBLICATIONS PVT. LTD.
NEW DELHI - 110 002 (INDIA)

ANMOL PUBLICATIONS PVT. LTD.

H.O.: 4374/4B, Ansari Road, Darya Ganj,
New Delhi-110 002 (India)
Ph.: 23278000, 23261597

B.O.: No. 1015, Ist Main Road, BSK IIIrd Stage
IIIrd Phase, IIIrd Block
Bangalore - 560 085 (India)
Visit us at: www.anmolpublications.com

Basics of Molecular Biology

ISBN 978-81-261-3501-1

PRINTED IN INDIA

Printed at Mehra Offset Press, Delhi.

Contents

Preface

Molecular Biology is one of the most rapidly moving areas of the life sciences, and a number of substantial advances have been made. Particularly noteworthy progress has been made in the area of genome sequencing: We now know the complete sequences not only of yeast and bacterial genomes but also of *C. elegans* and *Drosophila melanogaster*, as well as much of the genomic sequence of the model plant *Arabidopsis thaliana*. Projects directed toward sequencing the human genome have also moved forward rapidly, and completion of the human genome sequence can be confidently anticipated in the near future. These advances in genome sequencing are opening whole new approaches to understanding the regulation and function of cells, and they will undoubtedly provide a new database upon which much of the future research in molecular biology will be built.

There have been significant advances in the understanding of several aspects of the molecular biology of eukaryotic cells—the mechanisms that control the initiation of DNA replication have been elucidated, studies of histone acetylation have provided a clear link between chromatin structure and transcriptional regulation, and new mechanisms of translational regulation have been described. Substantial progress has similarly been made in understanding a variety of aspects of protein trafficking, including nuclear import and export, the pathways of protein import to the endoplasmic reticulum and transport to the Golgi apparatus, the molecular mechanisms of vesicular transport, and the mechanisms of protein import into mitochondria and chloroplasts. The exciting areas of cell signaling and cell cycle regulation have also continued to rapidly move forward, with particularly noteworthy progress being made in understanding the regulation of programmed cell death and the signaling pathways that control cell fate during embryonic development. These advances in basic science have been coupled with progress in medically-related areas, including the potential use of stem cells to replace damaged human tissue and the development of new anticancer drugs targeted against specific oncogene proteins.

Author

Chapter 1

Introduction

Molecular biology is the study of biology at a molecular level. The field overlaps with other areas of biology and chemistry, particularly genetics and biochemistry. Molecular biology chiefly concerns itself with understanding the interactions between the various systems of a cell, including the interrelationship of DNA, RNA and protein synthesis and learning how these interactions are regulated.

Writing in *Nature,* William Astbury described molecular biology as:

"... not so much a technique as an approach, an approach from the viewpoint of the socalled basic sciences with the leading idea of searching below the large-scale manifestations of classical biology for the corresponding molecular plan. It is concerned particularly with the forms of biological molecules and is predominantly three dimensional and structural - which does not mean, however, that it is merely a refinement of morphology - it must at the same time inquire into genesis and function."

RELATIONSHIP TO OTHER "MOLECULAR-SCALE" BIOLOGICAL SCIENCES

Researchers in molecular biology use specific techniques native to molecular biology, but increasingly combine these with techniques and ideas from genetics and biochemistry. There is not a hard-line between these disciplines as there once was. The following figure is a schematic that depicts one possible view of the relationship between the fields:

- *Biochemistry* is the study of the chemical substances and vital processes occurring in living organisms.
- *Genetics* is the study of the effect of genetic differences on organisms. Often this can be inferred by the absence of a normal component (e.g. one gene). The study of "mutants" – organisms which lack one or more functional components with respect to the so-called "wild type" or normal phenotype. Genetic interactions such as epistasis can often confound simple interpretations of such "knock-out" studies.
- *Molecular biology* is the study of molecular underpinnings of the process of replication, transcription and translation of the genetic material. The central dogma of molecular biology where genetic material is transcribed into RNA and then translated into protein, despite being an oversimplified picture of molecular biology, still provides a good starting point for understanding the field. This picture, however, is undergoing revision in light of emerging novel roles for RNA.

Much of the work in molecular biology is quantitative, and recently much work has been done at the interface of molecular biology and computer science in bioinformatics and computational biology. As of the early 2000s, the study of gene structure and function, molecular genetics, has been amongst the most prominent sub-field of molecular biology.

Increasingly many other fields of biology focus on molecules, either directly studying their interactions in their own right such as in cell biology and developmental biology, or indirectly, where the techniques of molecular biology are used to infer historical attributes of populations or species, as in fields in evolutionary biology such as population genetics and phylogenetics. There is also a long tradition of studying biomolecules "from the ground up" in biophysics.

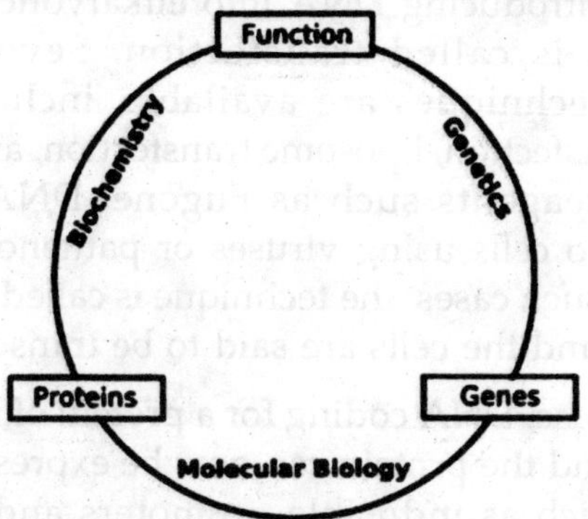

Fig. Schematic Relationship Between Biochemistry, Genetics and Molecular Biology

TECHNIQUES OF MOLECULAR BIOLOGY

Since the late 1950s and early 1960s, molecular biologists have learned to characterize, isolate, and manipulate the molecular components of cells and organisms. These components include DNA, the repository of genetic information; RNA, a close relative of DNA whose functions range from serving as a temporary working copy of DNA to actual structural and enzymatic functions as well as a functional and structural part of the translational apparatus; and proteins, the major structural and enzymatic type of molecule in cells.

EXPRESSION CLONING

One of the most basic techniques of molecular biology to study protein function is expression cloning. In this technique, DNA coding for a protein of interest is cloned (using PCR and/ or restriction enzymes) into a plasmid (known as an expression vector). This plasmid may have special promoter elements to drive production of the protein of interest, and may also have antibiotic resistance markers to help follow the plasmid.

This plasmid can be inserted into either bacterial or animal cells. Introducing DNA into bacterial cells is called transformation, and can be completed with several methods, including electroporation, microinjection, passive uptake and conjugation. Introducing DNA into eukaryotic cells, such as animal cells, is called transfection. Several different transfection techniques are available, including calcium phosphate transfection, liposome transfection, and proprietary transfection reagents such as Fugene. DNA can also be introduced into cells using viruses or pathenogenic bacteria as carriers. In such cases, the technique is called viral/bacterial transduction, and the cells are said to be transduced.

In either case, DNA coding for a protein of interest is now inside a cell, and the protein can now be expressed. A variety of systems, such as inducible promoters and specific cell-signaling factors, are available to help express the protein of interest at high levels. Large quantities of a protein can then be extracted from the bacterial or eukaryotic cell. The protein can be tested for enzymatic activity under a variety of situations, the protein may be crystallized so its tertiary structure can be studied, or, in the pharmaceutical industry, the activity of new drugs against the protein can be studied.

POLYMERASE CHAIN REACTION (PCR)

The polymerase chain reaction is an extremely versatile technique for copying DNA. In brief, PCR allows a single DNA sequence to be copied (millions of times), or altered in predetermined ways. For example, PCR can be used to introduce restriction enzyme sites, or to mutate (change) particular bases of DNA. PCR can also be used to determine whether a particular DNA fragment is found in a cDNA library. PCR has many variations, like reverse transcription PCR (RT-PCR) for amplification of RNA, and, more recently, real-time PCR (qPCR) which allow for quantitative measurement of DNA or RNA molecules.

GEL ELECTROPHORESIS

Gel electrophoresis is one of the principal tools of molecular biology. The basic principle is that DNA, RNA, and proteins can all be separated using an electric field. In agarose gel electrophoresis, DNA and RNA can be separated based on size by running the DNA through an agarose gel. Proteins can be separated based on size using an SDS-PAGE gel. Proteins can also be separated based on their electric charge, using what is known as an isoelectric gel.

SOUTHERN BLOTTING

Named after its inventor, biologist Edwin Southern, the Southern blot is a method for probing for the presence of a specific DNA sequence within a DNA sample. DNA samples before or after restriction enzymedigestion are separated by gel electrophoresis and then transferred to a membrane by blotting via capilliary action. The membrane can then be probed using a DNA probe labeled using a complement of the sequence of interest. Most original protocols used radioactive labels, however now non-radioactive alternatives are available. Southern blotting is less commonly used in laboratory science due to the capacity of using PCR to detect specific DNA sequences from DNA samples. However, these blots are still used for some applications, such as measuring transgene copy number in transgenic mice, or in the engineering of gene knockout embryonic stem cell lines.

NORTHERN BLOTTING

The Northern blot is used to study the expression patterns a specific type of RNA molecule as relative comparison among of a set of different samples of RNA. It is essentially a combination of denaturing RNA gel electrophoresis, and a blot. In this process RNA is separated based on size and is then transferred to a membrane that is then probed with a labeled complement of a sequence of interest. The results may be visualized through a variety of ways depending on the label used, however, most result in the revelation of bands representing the sizes of the RNA detected in sample. The intensity of these bands is related to the amount of the target RNA in the samples analyzed. The procedure is commonly used to study when and how much gene expression is occurring by measuring how much of that RNA is present in different samples. It is one of the most basic tools for determining at what time certain genes are expressed in living tissues.

WESTERN BLOTTING

Antibodies to most proteins can be created by injecting small amounts of the protein into an animal such as a mouse, rabbit, sheep, or donkey (polyclonal antibodies)or produced in cell culture (monoclonal antibodies). These antibodies can be used for a variety of analytical and preparative techniques.

In Western blotting, proteins are first separated by size, in a thin gel sandwiched between two glass plates in a technique known as SDS-PAGE (for Sodium Dodecyl Sulphate Poly-Acrylamide Gel Electrophoresis). The proteins in the gel are then transferred to a PVDF, nitrocellulose, nylon or other support membrane. This membrane can then be probed with solutions of antibodies. Antibodies that specifically bind to the protein of interest can then be visualized by a variety of techniques, including coloured products, chemiluminescence, or autoradiography.

Analogous methods to western blotting can also be used to directly stain specific proteins in cells and tissue sections. However, these *immunostaining* methods are typically more associated with cell biology than molecular biology.

The terms "Western" and "Northern" are jokes: The first blots were with DNA, and since they were done by Ed Southern, they came to be known as Southerns. I don't think Patricia Thomas, inventor of the RNA blot, which became known as a "Northern", actually used the term.To carry the joke further, one can find reference in the literature to "southwesterns" (Protein-DNA interactions) and "far-westerns" (Protein-Protein interactions).

ARRAYS

A DNA array is a collection of spots attached to a solid support such as a microscope slide; each spot contains one or more DNA molecules. Arrays make it possible to put down a large number of very small (100 micrometre diameter) spots on a single slide; if each spot has a DNA molecule that is complementary to a single gene, one can analyze the expression of every gene in an organism in a single experiment. For instance, the common baker's yeast, *Saccharomyces cerevisiae,* contains about 7000 genes; with a microarrray, one can measure quantitatively, how each gene is expressed, and how that expression changes, for example, with a change in temperature. There are many different ways to fabricate microarrays; the most common are silicon chips, microscope slides with spots of ~ 100 micrometre diameter, custom arrays, and arrays with larger spots on porous membranes (macroarrays).

Arrays can also be made with molecules other than DNA. For example, an antibody array can be used to determine what proteins or bacteria are present in a blood sample

ABANDONED TECHNOLOGY

As new procedures and technology become available, the older technology is rapidly abandoned. A good example is methods for determining the size of DNA molecules. Prior to gel electrophoresis, with agarose and polyacrylamide, DNA was sized with rate sedimentation in sucrose gradients, a slow and labour intensive technology requiring expensive instrumentation; prior to sucrose gradients, viscometry was used.

Aside from their historical interest, it is worth knowing about older technology as it may be useful to solve a particular problem.

HISTORY

Molecular biology was established in the 1930s, the term was first coined by Warren Weaver in 1938 however. Warren was director of Natural Sciences for the Rockefeller Foundation at the time and believed that biology was about to undergo a period of significant change given recent advances in fields such as X-ray crystallography. He therefore channeled significant amounts of (Rockefeller Institute) money into biological fields.

Chapter 2

Protein: Structure and Functions

BUILDING BLOCKS - AMINO ACIDS

An amino acid is defined as the molecule containing an amino group (NH_2), a carboxyl group (COOH) and an R group. It has the following general formula,

$R\text{-}CH(NH_2)\text{-}COOH$

The R group differs among various amino acids. In a protein, the R group is also called a sidechain.

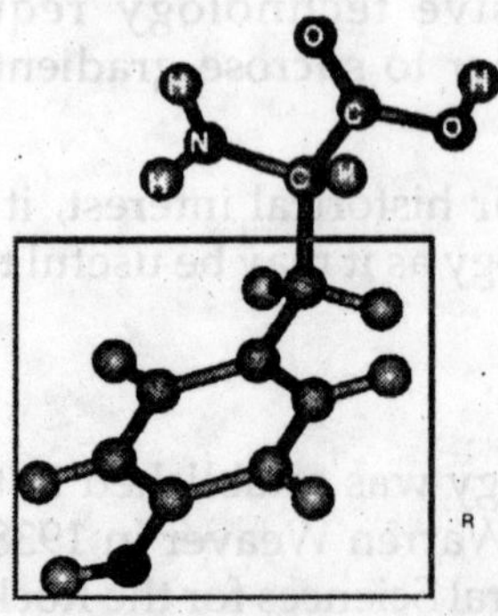

Fig. The Structure of an Amino Acid (Tyrosine).

There are over 300 naturally occurring amino acids on earth, but the number of different amino acids in proteins is only 20.

THE TWENTY AMINO ACIDS OF PROTEINS

Based on the physical and chemical properties of R groups, the 20 amino acids of proteins may be classified as follows.

1. Acidic: including aspartic acid (aspartate) and glutamatic acid (glutamate). In a neutral solution, the R group of an acidic amino acid may lose a proton and become negatively charged.
2. Basic: including lysine, arginine and histidine. In a neutral solution, the R group of a basic amino acid may gain a proton and become positively charged. Interaction between positive and negative R groups may form a salt bridge, which is an important stabilizing force in proteins.
3. Aromatic: including tyrosine, tryptophan and phenylalanine. Their R groups contain an aromatic ring.
4. Sulfur: including cysteine and methionine. Their R groups contain a sulfur atom (S). The disulfide bond formed between two cysteine residues provides a strong force for stabilizing the globular structure. A unique feature about methionine is that the synthesis of all peptide chains starts from methionine.
5. Uncharged hydrophilic: including serine, threonine, asparagine and glutamine. Their R groups are hydrophilic and capable of forming hydrogen bonds.
6. Inactive hydrophobic: including glycine, alanine, valine, leucine and isoleucine. These amino acids are more likely to be buried in the protein interior. Their R groups do not form hydrogen bonds and rarely participate in chemical reactions.
7. Special structure: including proline. In most amino acids, the R group and the amino group are not directly connected. Proline is the only exception among 20 amino acids found in protein. Due to this special feature, proline is often located at the turn of

a peptide chain in the three-dimensional structure of a protein.

PH AND PKA

A molecule, or an atom group in a molecule, may lose or gain a proton when the molecule is placed in an aqueous solution. The exact probability that a molecule will be protonated or deprotonated depends on the pK_a of the molecule and the pH of the solution.

Let AH be an atom group in a molecule. AH could be neutral or charged. After AH loses a proton, it is denoted by A^-. The protonation/deprotonation reaction may be written as

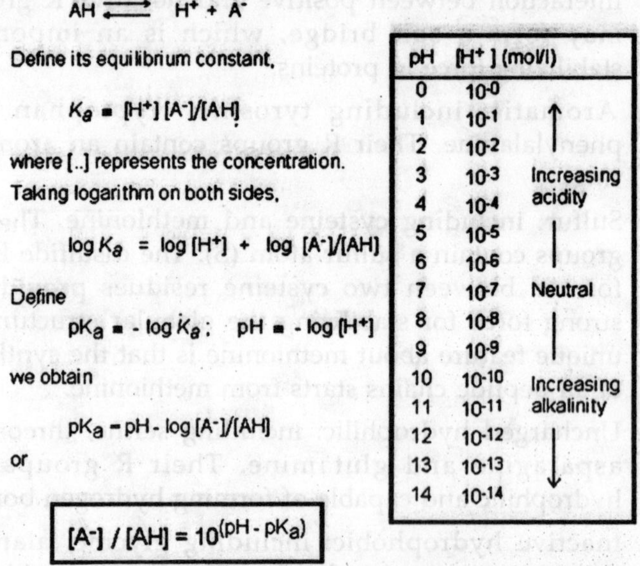

$$AH \rightleftharpoons H^+ + A^-$$

Define its equilibrium constant,

$$K_a = [H^+][A^-]/[AH]$$

where [..] represents the concentration.

Taking logarithm on both sides,

$$\log K_a = \log [H^+] + \log [A^-]/[AH]$$

Define

$$pK_a = -\log K_a; \quad pH = -\log [H^+]$$

we obtain

$$pK_a = pH - \log [A^-]/[AH]$$

or

$$[A^-] / [AH] = 10^{(pH - pK_a)}$$

pH	$[H^+]$ (mol/l)	
0	10^{-0}	↑
1	10^{-1}	
2	10^{-2}	
3	10^{-3}	Increasing acidity
4	10^{-4}	
5	10^{-5}	
6	10^{-6}	
7	10^{-7}	Neutral
8	10^{-8}	
9	10^{-9}	
10	10^{-10}	Increasing alkalinity
11	10^{-11}	
12	10^{-12}	
13	10^{-13}	
14	10^{-14}	↓

From the last equation, we can see that half of molecules will lose protons if they are in a solution with $pH = pK_a$. The higher the pH value, the more likely a molecule will lose a proton. On the other hand, from the definition of pH, we find that the proton concentration of a solution becomes lower at higher pH, thereby capable of accepting more protons. This physical concept is consistent with the result we just obtained from the equation.

- The carboxyl group is most likely negatively charged.
- The amino group is most likely positively charged
- The R group of aspartate and glutamate are most likely negatively charged.
- The R group of lysine and arginine are most likely positively charged.
- The R group of tyrosine at pH = 7 is most likely neutral.
- The R group of histidine has 10per cent probability to become positively charged at pH = 7, but the probability increases to 50per cent at pH = 6.

  Thus, histidine is very sensitive to pH change in the physiological range.

Protonation/Deprotonation reaction	pK_a
$HO-P(=O)(OH)-OH \rightleftharpoons {}^{-}O-P(=O)(OH)-OH + H^+$	2
${}^{-}O-P(=O)(OH)-OH \rightleftharpoons {}^{-}O-P(=O)(O^-)-OH + H^+$	7
${}^{-}O-P(=O)(O^-)-OH \rightleftharpoons {}^{-}O-P(=O)(O^-)-O^- + H^+$	12

Fig. pK_a Values of Phosphate Groups.

This information tells us that in a neutral solution H_3PO_4 is very likely to lose a proton and become H_2PO4^-. The latter has a 50per cent chance of losing another proton, becoming HPO_4^{2-}. The chance for HPO_4^{2-} to lose an additional proton is very small. Hence, the average charge of a phosphate group is about 1.5 electron charges.

PEPTIDES

A peptide is a chain of amino acids linked together by peptide bonds. Polypeptides usually refer to long peptides whereas oligopeptides are short peptides (< 10 amino acids). Proteins are made up of one or more polypeptides with more than 50 amino acids.

PRIMARY STRUCTURE

The primary structure of a protein refers to its amino acid sequence. The amino acid in a peptide is also called a residue.

1 ... 20

NH_3^+—KETAAAKFERQHMDSSTSAASSSNYC
N
Q
M
VAQVDALSEHVFTNVPKCRDKTLNRSKM
58C SQKNVACKNGQTNCYQSYSTMSITDCRETGS
S 90
COO^-—VSADFHVPVYPNGECAVIIHKNAQTTKYACNPYK
124

Fig: The Amino Acid Sequence (Primary Structure) of Ribonuclease A (RNase A), which is an Enzyme Acting on RNA. Each Letter Represents an Amino Acid.

PEPTIDE BOND

A peptide bond is the linkage between two amino acids, formed by the condensation reaction, as shown below:

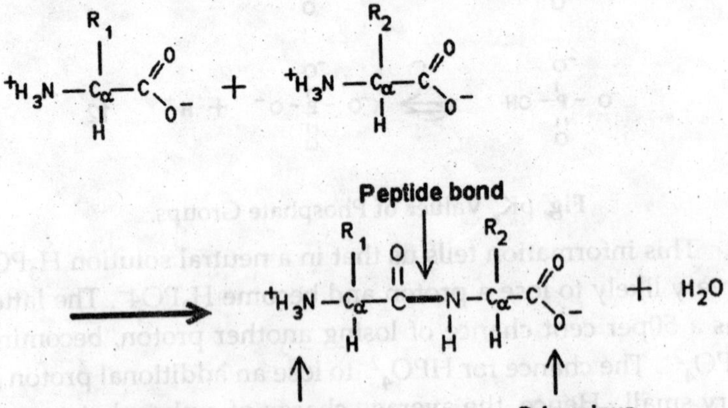

$$^{+}H_3N-C_{\alpha}H(R_1)-COO^{-} + {}^{+}H_3N-C_{\alpha}H(R_2)-COO^{-}$$

$$\longrightarrow {}^{+}H_3N-C_{\alpha}H(R_1)-C(=O)-NH-C_{\alpha}H(R_2)-COO^{-} + H_2O$$

Peptide bond

N-terminus

C-terminus

Fig. Formation of the Peptide Bond by Condensation Reaction

PHI-PSI ANGLES

Due to the specific electronic structure of the peptide bond, the atoms on its two ends cannot rotate around the bond. Hence, the atoms of the group, O=C–N–H, are fixed on the same plane, known as the peptide plane. The whole plane may rotate around the N-C_α bond (ϕ angle) or C-C_α bond (Ψ angle). C_α is the carbon atom connected to the R group.

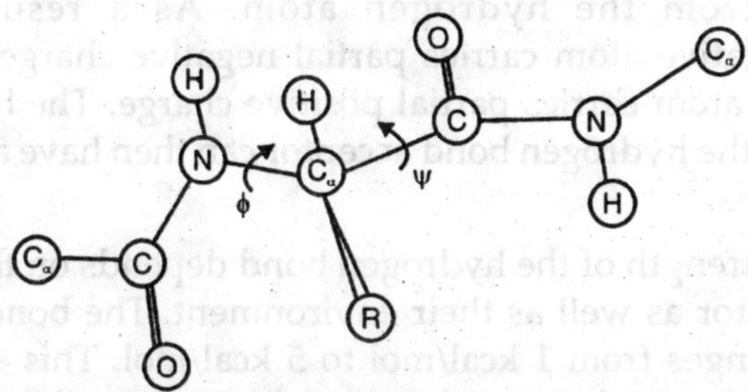

Fig. Illustration of the Peptide Plane (Gray Area) and f-Y Angles. The Red Line formed by the Repeating –C_a–C–N–C_a– is the Backbone of the Peptide Chain.

Mathematically, phi (ϕ) and psi (Ψ) are the dihedral angle (also known as torsional angle) which is defined as the angle between the point (e.g., C_α) at the end of a 4-point sequence and the plane (e.g., peptide plane) occupied by the other three points. In a peptide, phi-psi angles are restricted to certain ranges. A plot of their distribution is called the Ramachandran plot.

SECONDARY STRUCTURE

In a protein, certain domains may form specific structures such as α helix and bstrand, which constitute the secondary structure of the protein.

```
1    KETAAAKFER QHMDSSTSAA SSSNYCNQMM KSRNLTKDRC KPVNTFVHES
          α1                      α2                   β1

51   LADVQAVCSQ KNVACKNGQT NCYQSYSTMS ITDCRETGSS KYPNCAYKTT
         α3         β2         β3         β4         β5

101  QANKHIIVAC EGNPYVPVHF DASV
             β6              β7
```

Fig. The Secondary Structure of RNase A, which Contains Three a Helices and Seven b Strands.

HYDROGEN BOND

A hydrogen bond is formed by three atoms: one hydrogen atom and two electronegative atoms (often N or O). The hydrogen atom is covalently bonded to one of electronegative atoms, called the hydrogen bond donor. The other electronegative atom is named the hydrogen bond acceptor. The two electronegative atoms may take up some electron density from the hydrogen atom. As a result, each electronegative atom carries partial negative charge and the hydrogen atom carries partial positive charge. The hydrogen atom and the hydrogen bond acceptor can then have attractive interactions.

The strength of the hydrogen bond depends on the donor and acceptor as well as their environment. The bond energy usually ranges from 1 kcal/mol to 5 kcal/mol. This energy is smaller than covalent bond energy, but greater than thermal energy (0.6 kcal/mol at room temperature). Therefore, the hydrogen bond can provide a significant stabilizing force in macromolecules such as proteins and nucleic acids.

When a hydrogen bond is formed, the distance between the donor and the acceptor is about 0.28 nm (1 nm = 10^{-9} meter). The hydrogen atom is still covalently bonded to the donor and their distance is shorter than that between the hydrogen atom and the acceptor.

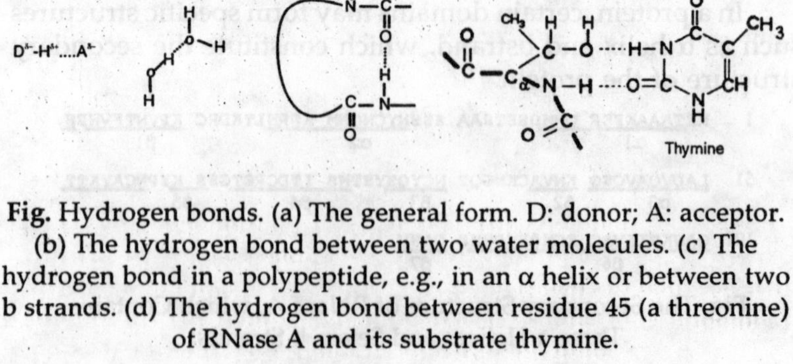

(a) $D^{\delta-}$-$H^{\delta+}$....$A^{\delta-}$

(b) (c) (d) Ribonuclease A (RNase A) T45 Thymine

Fig. Hydrogen bonds. (a) The general form. D: donor; A: acceptor. (b) The hydrogen bond between two water molecules. (c) The hydrogen bond in a polypeptide, e.g., in an α helix or between two b strands. (d) The hydrogen bond between residue 45 (a threonine) of RNase A and its substrate thymine.

ALPHA HELIX

An α helix has the following features:

- Every 3.6 residues make one turn,
- The distance between two turns is 0.54 nm,
- The C=O (or N-H) of one turn is hydrogen bonded to N-H (or C=O) of the neighboring turn.

Hydrogen bonds play a role in stabilizing the α helix conformation. However, the size and charges of sidechains are also important factors. Alanine has a greater propensity to form α helices than proline.

An α helix can be either right-handed or left-handed, as defined in the following figure.

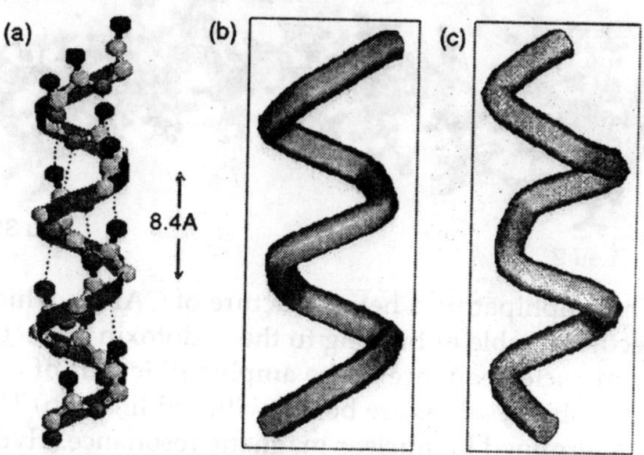

Fig. a Helix Conformations. (a) The Ideal Right-handed a helix. C: Green; O: Red; N: Blue; H: not shown; Hydrogen Bond: dashed line. (b) The right-handed a helix without showing atoms. (c) the left-handed a helix.

The right-handed or left-handed helix may be distinguished by stretching out your thumb and curling the other four fingers. Imaging that in the above figure the helix is spiraling upward. Stretch your right thumb upward, then the other four fingers of the right hand will be able to curl in the same direction as each turn in the spiral. The same result can be obtained if you imagine that the helix is spiraling downward and point your thumb downward. The left hand works for the left-handed α helix.

AMPHIPATHIC ALPHA HELIX

In an amphipathic α helix, one side of the helix contains mainly hydrophilic amino acids and the other side contains mainly hydrophobic amino acids. The amino acid sequence of amphipathic α helix alternates between hydrophilic and hydrophobic residues every 3 to 4 residues, since the α helix makes a turn for every 3.6 residues. An example is shown below.

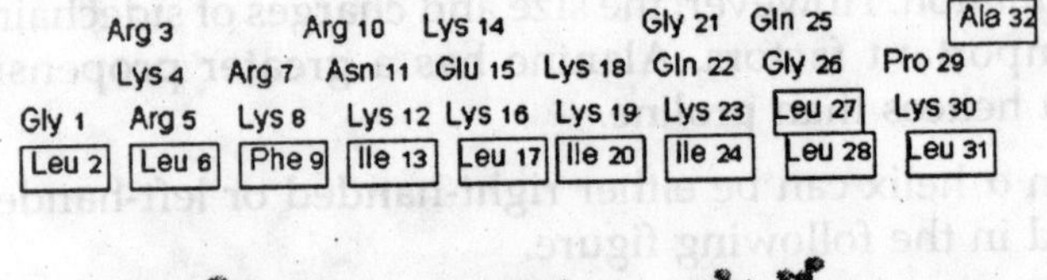

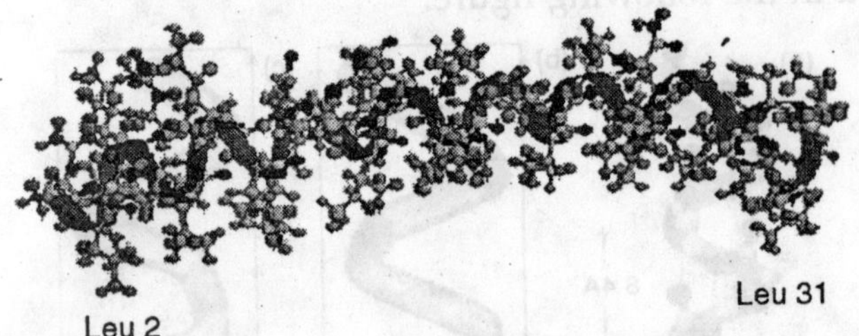

Fig. The amphipathic α helix structure of CAP18, which is a molecule capable of binding to the endotoxin of bacteria. (a) Amino acid sequence of the amphipathic part of CAP18. Hydrophobic residues are boxed with red lines. (b) The 3D structure determined by nuclear magnetic resonance. Hydrophobic residues are located on the lower side. PDB ID = 1LYP.

BETA STRAND, BETA SHEET AND BETA BARREL

Beta Strand

In a β strand, the torsion angle of N-C_α-C-N in the backbone is about 120 degrees. The following figure shows the conformation of an ideal β strand. Note that the sidechains of two neighboring residues project in the opposite direction from the backbone.

Beta Sheet

A β sheet consists of two or more hydrogen bonded β strands. The two neighboring b strands may be parallel if they are aligned in the same direction from one terminus (N or C) to the other, or anti-parallel if they are aligned in the opposite direction.

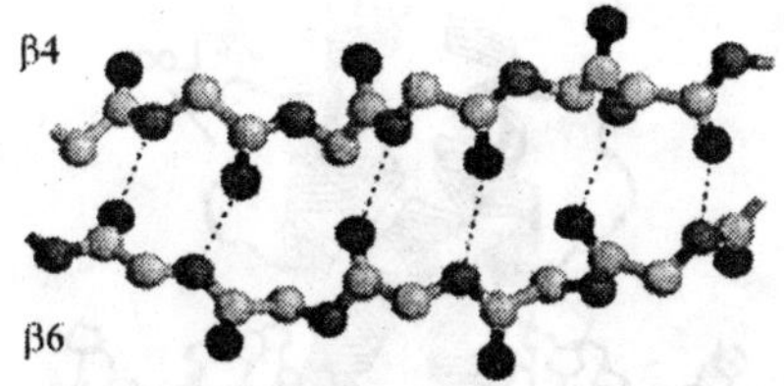

Fig. The β sheet structure found in RNase A.

This figure shows only the backbone atoms, excluding hydrogens. RNase A contains a single peptide chain, which makes a turn at the junction (not shown) between β4 and β6. Therefore, the two strands are antiparallel.

A β barrel is a closed β sheet.

PROTEIN MOTIFS AND DOMAINS

The motif is a characteristic domain structure consisting of two or more a helices or β strands. Common examples include coiled coil, helix-loop-helix, zinc finger, leucine zipper, etc.

Many proteins also contain specific domains such as the SH2 domain.

Helix-Loop-Helix

"Helix-Loop-Helix" is characterized by two a helices connected by a loop. It should not be confused with "Helix-Turn-Helix". In fact, these two structures are quite different, but the Helix-Loop-Helix looks like leucine zipper. Example: myoblast determination proteins (MyoD).

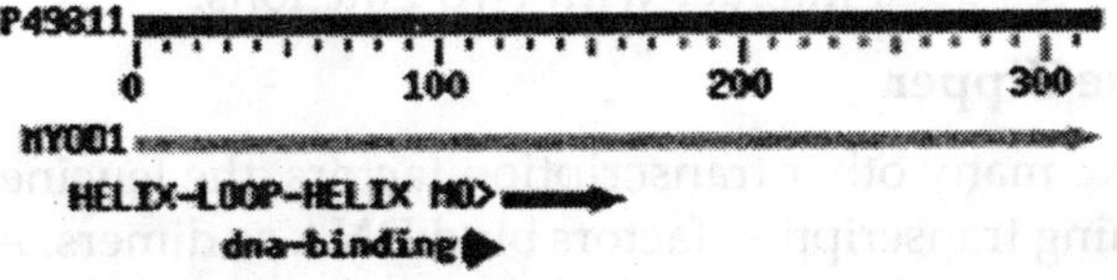

Fig. The transcription factor MyoD. (a) The domain organization. (b) The structure of the helix-loop-helix motif. PDB ID = 1MDY.

Zinc Finger

The zinc finger motif contains one or more zinc ions which are crucial for the structural stability. It can be divided into three types:

C_2H_2 zinc finger: It is characterized by the sequence $CX_{2-4}C....HX_{2-4}H$, where C = cysteine, H = histidine, X = any amino acid. In the 3D structure, two cysteine residues and two histidine residues interact with a zinc ion.

C_4 zinc finger: Its consensus sequence is $CX_2CX_{13}CX_2CX_{14-15}CX_5CX_9CX_2C$. The first four cysteine residues bind to a zinc ion and the last four cysteine residues bind to another zinc ion.

C_6 zinc finger. It has the consensus sequence $CX_2CX_6CX_{5-6}CX_2CX_6C$. The yeast's Gal4 contains such a motif where six cysteine residues interact with two zinc ions.

Leucine Zipper

Like many other transcription factors, the leucine-zipper-containing transcription factors bind DNA as dimers. A leucine zipper is formed by two α helices, one from each monomer. The helices are held together by hydrophobic interactions between leucine residues, which are located on one side of each helix. Examples include AP-1, CREB, and Gcn4.

THREE-DIMENSIONAL STRUCTURE

The three-dimensional (3D) structure is also called the tertiary structure. If a protein molecule consists of more than one polypeptide, it also has the quaternary structure, which specifies the relative positions among the polypeptides (subunits) in a protein.

Fig. The Ribbon Representation of the 3D Structure of RNase A. Its Atomic Coordinates were obtained from Protein Data Bank (PDB ID = 1RCN).

X-ray crystallography and nuclear magnetic resonance (NMR) are the major experimental techniques for determining the 3D structures of macromolecules. Protein Data Bank is the most important site for structural information which can easily be accessed by the "PDB ID" of a macromolecule.

ENZYMES

Enzymes are the catalysts of biochemical reactions in the cell. Numerous drugs are enzyme inhibitors.

Prominent enzymes as drug targets:

- HIV Protease: Essential for HIV replication. Its inhibitor has been able to control AIDS.
- Angiotensin Converting Enzyme (ACE): Promotes contraction of blood vessels. Its inhibitors have been widely used to treat high blood pressure and congestive heart failure.
- HMG-CoA Reductase: Essential for the synthesis of cholesterol. Its inhibitors (such as statins) can reduce the cholesterol level.
- cGMP Phosphodiesterase: Catalyzes the conversion of cGMP into GMP. It is the target of the well known drug for treating impotence, Viagra.
- Calcineurin: A phosphatase that plays an important role in immune responses. It is the target of the immunosuppressive drugs FK506 and cyclosporin A.

ENZYME INHIBITORS

Numerous drugs are enzyme inhibitors which may be divided into two types: competitive inhibitors and non-competitive inhibitors. The competitive inhibitor directly occupy the active site before the substrate can bind to it. The non-competitive inhibitor does not occupy the active site, but its binding to other region somehow inhibits the catalytic activity.

HIV protease inhibitor is a competitive inhibitor.

Angiotensin II

We are talking about the chemical messengers in the body called neurohormones. For those of us with heart failure, the neurohormone called angiotensin II is not a good thing.

Angiotensin II helps controls the cardiovascular system. It causes blood vessels to shrink (vasoconstriction), which raises blood pressure. Angiotensin II also causes the body to release aldosterone - a substance that causes our kidneys to retain sodium and fluid, causing edema.

Angiotensin II also causes the body to release a substance called vasopressin or ADH, which causes us to retain fluid even more. Angiotensin II is a real bug-a-boo for us CHFers.

ACE Inhibitors

We all have angiotensin one in our systems. Angiotensin one is converted into angiotensin II by an enzyme in your body called ACE (Angiotensin Converting Enzyme). ACE is what we block with ACE inhibitors - blocking it prevents angiotensin I from converting into angiotensin II.

Reducing angiotensin II lets our blood vessels relax and expand. This lowers blood pressure, which makes the heart's job a lot easier.

Think of it this way: If you pump the same amount of water at the same speed through a little pipe as through a big pipe, there's higher pressure in the small one. The fluid *shoots* out of the small pipe, but calmly flows out of the big one, even though they are both moving the same amount of water. The pump pushing the water through that skinny pipe has to work a lot harder than the pump glushing it through the big pipe.

The same principle applies to our blood vessels. If our arteries relax, they get bigger (vasodilation) and the pump has an easier job - less pressure. That's called reducing "afterload." ACE inhibitors do this for us. "Super" ACE inhibitors called vasopeptidase inhibitors were also tested for CHF but carry higher risk for side effects.

ACE also helps control the amount of blood in your body - your blood "volume." Blocking ACE reduces the overall amount of blood in your body. This means your tired old heart has less blood to pump, which makes its job easier. ACE inhibitors increase the level of potassium in your blood, so they partly offset potassium loss caused by diuretics like Lasix and Demadex.

Finally, ACE inhibitors help us live longer and feel better. The articles below show that ACE inhibitors reduce our risk of death by 20per cent to 40per cent. The other two types of drugs that do this are beta-blockers and Aldactone (spironolactone). ACE inhibitors slow heart remodeling, preventing our hearts from getting even weaker over time.

It can take several months for the full effect of ACE inhibitors to show up, and improve our quality of life. Hang in there, because ACE inhibitors *do* work and *all* CHFers should be on them if possible. It's even in the official heart failure treatment guidelines.

Unfortunately, many doctors don't offer us ACE inhibitors if we have any kidney problems or if our blood pressure is low. Monopril (fosinopril) is processed by the liver as well as the kidneys, so CHFers with kidney problems can usually take it okay. People with low blood pressure, surprisingly enough, often tolerate ACE inhibitors quite well (although not always).

Angiotensin II Receptor Blockers[ARBs]

Angiotensin II is also made at other places in the body. So ACE inhibitors can't *completely* stop angiotensin II from affecting you. After starting an ACE inhibitor, there is a period where angiotensin II is almost completely stopped. Over time, your angiotensin II level rises again, from these other places in the body. Since ACE inhibitors can't do it all, it seems like a good idea to block the stuff at other places too. ARBs block angiotensin II receptors on cell walls. If angiotensin II can't connect to cell receptors, it can't affect us. The most important receptor is called the AT-1 receptor (there is also an AT-2 receptor). ARBs *do* help CHF. The question is: Do ARBs help CHFers already taking an ACE inhibitor? This has not been completely decided yet.

The resolved study compared an ARB alone to using an ACE inhibitor alone to using both an ACE inhibitor and an ARB. Using both worked better to improve heart enlargement and EF.

The ValHeft trial added an ARB to standard CHF therapy. Complications and mortality improved most in CHFers taking the ARB with either an ACE inhibitor *or* a beta-blocker, but not in those taking all 3 drugs. Why taking all 3 drugs gave worse results is unknown.

ARBs do have some excellent uses we *are* sure about. In patients who develop a bad cough from using an ACE inhibitor, an ARB can often be used without the cough. In patients who don't tolerate an ACE inhibitor for some other reason, an ARB will often do the job.

Note- Telmisartan can significantly raise blood digoxin levels.

Conclusion

High doses are better than low doses but doctors don't usually prescribe them - due to simple ignorance. Double-check your doc on this to be sure you are getting a high enough dose!

Even if you improve to normal heart function and consider yourself "cured," a good CHF specialist will keep you on an ACE inhibitor for life, and that's smart. It may stop you from getting CHF again down the road - which can happen.

ACE inhibitor use may reduce your zinc level so you might want to take a zinc supplement. Your sense of taste could possibly go goofy on you after starting an starting ACE inhibitor too - especially on captopril.

Synthesis of Lipids

Lipid synthesis occurs at the smooth endoplasmic reticulum. Lipids may be transported to appropriate locations by vesicles or by phospholipid exchange proteins.

Fueled by fat-rich diets and sedentary lifestyles, atherosclerosis and obesity are major health issues in the Western world. Both diseases involve excessive accumulation of neutral lipids: cholesterol esters in arteries in atherosclerosis and triglycerides in adipocytes in obesity. The study of these diseases has sparked interest in the biochemistry and molecular biology of neutral lipid synthesis.

The discovery of acyl-CoA:cholesterol acyltransferase (ACAT,[1] EC 2.3.1.26), a cholesterol ester synthesis enzyme, dates back to the 1950s. ACAT was identified as an intracellular enzyme in the endoplasmic reticulum (ER) that covalently joined cholesterol and fatty acyl-CoA molecules to form cholesterol esters. ACAT activity was detected in macrophages and in tissues such as liver, small intestine, and adrenal glands. Another ER enzyme, acyl-CoA:diacylglycerol acyltransferase (DGAT, EC 2.3.1.20), was identified in 1960. The DGAT reaction is similar to that of ACAT except that diacylglycerol is the acyl group acceptor. In the ensuing 30 years, much was learned about the biochemistry of ACAT and DGAT. Both enzymes were found to play important roles in synthesizing neutral lipids for the assembly and secretion of lipoproteins, and ACAT was found to be responsible for synthesizing cholesterol esters in arterial macrophage foam cells. However, these hydrophobic proteins proved difficult to isolate, slowing progress in their understanding.

Because of the pioneering work of Chang and colleagues the field of neutral lipid synthesis made great strides during the past decade. These investigators isolated ACAT by using a clever cloning strategy to isolate an ACAT cDNA. Thanks to the resultant molecular probes and to genetic studies in yeast and mice, we now know there are two mammalian ACAT enzymes and probably more than one DGAT enzyme. A large number of ACAT- and DGAT-related genes are identifiable in species ranging from plants to yeast to humans. Here we review recent advances in understanding the function of the mammalian members of this family, ACAT1, ACAT2, and DGAT.

GENE AND PROTEIN STRUCTURES

The human ACAT1 gene encodes four mRNAs of 7.0, 4.3, 3.6, and 2.8 kilobases (kb) (12, 18-20); all contain the same translational reading frame but differ in the length of the untranslated regions. The two shorter mRNAs are products of a proximal ACAT promoter. The 4.3-kb mRNA is derived from an unusual RNA recombination mechanism involving *trans*-splicing of two discontinuous precursor RNAs produced from chromosomes 1 and 7 (21). The origin of the 7.0-kb transcript is unknown. The human ACAT2 gene encodes a single mRNA of 2.2 kb, and the human DGAT gene encodes two mRNAs of 2.0 and 2.4 kb. Regions of shared sequence identity are present throughout the ACAT and DGAT proteins, but the greatest similarity is found in their C termini. Human and mouse ACAT2 are 40per cent identical to human ACAT1, and DGAT is ~20per cent identical to ACAT1.

Little is known about the relationship between protein structure and function in ACAT/DGAT family members. One motif, FYXDWWN (amino acids 403-409 of human ACAT1), is highly conserved in all family members and may be involved in fatty acyl-CoA binding. Another motif, MKXXSF (amino acids 265-270 of human ACAT1), is conserved in ACAT family members. The serine of this motif is required for ACAT activity. All three family members have a potential tyrosine phosphorylation motif and at least one *N*-linked glycosylation site, both of unknown significance. ACAT1 and ACAT2 contain an N-terminal leucine zipper motif. Although ACAT1 functions as a homotetramer it is unknown whether this leucine zipper motif participates in tetramer formation or whether ACAT2 and DGAT act as multimers. ACAT1 and ACAT2 do not form hetero-oligomeric complexes.

All ACAT/DGAT proteins possess multiple hydrophobic regions predicted to serve as transmembrane domains, although there is not yet a consensus structural model for the proteins. Investigations of human ACAT1 topology have yielded different results. One study, performed by inserting epitopes at different positions in the expressed protein found seven transmembrane domains. Another study, performed by expressing truncated forms of the protein found five transmembrane domains. In both studies, the N terminus of the protein localized to the cytoplasm and the C terminus to the ER lumen. Topologic studies of human ACAT2 performed with the truncation mutant method also found five transmembrane domains with the N terminus in the cytoplasm and the C terminus in the ER lumen. The latter study found that a serine residue essential for ACAT activity was located in the cytoplasm for ACAT1 and in the ER lumen for ACAT2, suggesting that the active sites of these enzymes reside on opposite sides of the ER membrane. Several putative transmembrane sequences are conserved in ACAT1 and ACAT2 but not in DGAT suggesting that these regions are involved in cholesterol binding. The topology of DGAT has not been reported, but a putative diacylglycerol binding motif has been identified.

TISSUE AND CELLULAR DISTRIBUTIONS

The tissue distributions of ACAT1 and ACAT2 are largely complementary. ACAT1 mRNA is present in many tissues, with the highest levels in macrophages and in adrenal and sebaceous glands all of which store cholesterol esters in cytoplasmic droplets. ACAT1 is also highly expressed in human atherosclerotic lesions, particularly in macrophage foam cells. ACAT2 is expressed predominantly in the liver and small intestine. In humans, nonhuman primates, and mice, ACAT2 appears to be the major ACAT in the small intestine and the predominant isozyme in hepatocytes of nonhuman primates and mice.

Whether ACAT1 or ACAT2 is the predominant enzyme in human hepatocytes is controversial. Both are expressed in human hepatocytes and HepG2 cells. However, immunodepletion experiments found that ACAT1 accounts for ~90per cent of ACAT activity in human liver, whereas ACAT2 accounts for most of the ACAT activity in adult small intestine and fetal liver but only 10-20per cent of activity in adult liver. It is unclear whether the hepatic ACAT1 activity resides in hepatocytes or macrophage-derived Kupffer cells. ACAT1 expression was detected immunohistochemically in both cell types, with stronger staining in Kupffer cells. In nonhuman primates, ACAT1 was expressed only in Kupffer cells in the liver. A significant proportion of the ACAT1 activity in human liver thus may be attributable to Kupffer cells. Another study found little evidence of ACAT2 expression in human hepatocytes.

DGAT mRNA and activity are ubiquitous in mouse and human tissues, with the highest levels in liver, small intestine, and adipose tissue and somewhat lower levels in testis and adrenal gland. This wide range of tissue expression is consistent with the involvement of DGAT in the glycerol phosphate pathway of triglyceride synthesis, which is common to most cells. DGAT protein expression in tissues has not been studied because of the lack of suitable antibodies.

The intracellular localization of enzymes in this class has been examined only for ACAT1. In human melanoma cells and fibroblasts and in mouse macrophages, ACAT1 is located primarily in the ER. In mouse macrophages, a small portion of ACAT1 immunoreactivity localizes to a region near the *trans*-Golgi network. In macrophages, ACAT1 localization may change with different conditions.

BIOCHEMICAL ACTIVITY AND REGULATION

ACAT and DGAT substrates have been identified by expressing the proteins in insect cells or yeast devoid of sterol esterification activity. In insect cells, mouse and human ACAT1 and ACAT2 utilize a variety of oxysterols in addition to cholesterol as substrates. Ergosterol and plant sterols are poor substrates. Both enzymes also utilize a wide variety of long chain fatty acyl-CoAs as substrates although arachidonyl-CoA appears to be a better substrate for ACAT1 than ACAT2. In insect cells, DGAT utilizes only diacylglycerol as the acyl acceptor. DGAT has a broad fatty acyl-CoA specificity.

ACAT1 is regulated primarily by post-translational mechanisms. In cholesterol-loaded cells, esterification increases without changes in the expression level of ACAT1 mRNA or protein. Purified ACAT1 is activated allosterically by cholesterol or oxysterols. ACAT1 activity may also be regulated by its intracellular localization. For example, a portion of ACAT1 immunoreactivity in macrophages is found near the *trans*-Golgi network and the endocytic recycling compartment. These cholesterol-rich organelles are involved in membrane traffic with the plasma membrane, and the proximity of ACAT1 to these organelles may be important for cholesterol ester synthesis during the formation of macrophage foam cells. The observation that 30-40per cent of ACAT1 redistributed from the ER to small vesicles during foam cell formation is consistent with this possibility. Regulation by phosphorylation has not been demonstrated.

ACAT1 is also transcriptionally regulated; however, only the expression levels of the shorter ACAT1 mRNAs (2.8 and 3.6 kb) change. In mice and rabbits a fat and cholesterol-rich diet increases ACAT1 mRNA levels 2-3-fold in the liver. In human HepG2 cells, free fatty acids, but not 25-hydroxycholesterol, increase ACAT1 mRNA levels 1.5-2-fold. ACAT1 mRNA and protein expression levels also increase during the differentiation of monocytes into macrophages. Two factors that induce gene expression during monocyte-macrophage differentiation1,25-dihydroxyvitamin D_3 and 9-*cis*-retinoic acid, also increased ACAT1 mRNA, protein, and cholesterol esterification activity in THP-1 cells.

Little is known about the regulation of ACAT2 and DGAT. Levels of DGAT mRNA increase markedly in parallel with enzyme activity in NIH 3T3-L1 cells during their differentiation into adipocytes. This finding suggests that transcription factors involved in adipogenesis activate DGAT expression. Like other family members, DGAT possesses a putative tyrosine phosphorylation site. The use of this site in DGAT regulation has not been demonstrated, although two reports have indicated that the enzyme may be posttranslationally regulated by a tyrosine kinase.

FUNCTIONAL ANALYSIS OF ACAT1 AND ACAT2

Gene knockout experiments in mice have provided insights into the biological roles of ACAT1 and ACAT2. In ACAT1-deficient mice, plasma cholesterol levels, hepatic cholesterol esterification activity, and intestinal cholesterol absorption are normal, consistent with the presence of a second ACAT enzyme in the small intestine and liver of mice. ACAT1 deficiency has its greatest effect in the adrenal glands and macrophages. Cholesterol esters and cholesterol esterification activity are virtually undetectable in the adrenal cortex of ACAT1$^{-/-}$ mice, although the depletion of cholesterol esters does not affect corticosterone synthesis. In addition, peritoneal macrophages from ACAT1$^{-/-}$mice do not accumulate cholesterol esters when incubated with cholesterol-containing lipoproteins. This finding provides a unique tool for testing the role of ACAT1 in macrophage foam cell formation and atherosclerosis. Another ACAT1 gene disruption study in mice confirmed general aspects of the ACAT1$^{-/-}$ phenotype and showed that ACAT1 deficiency reduces cholesterol esterification activity by >95per cent in testis and ovary and results in atrophy of the meibomian glands of the eyelids. A naturally occurring mutation in the mouse ACAT1 gene is present in the AKR strain of inbred mice which are homozygous for the *adrenocortical lipid depletion* (*ald*) allele.

The analysis of ACAT2-deficient mice (ACAT2$^{-/-}$) (34) complemented findings from ACAT1$^{-/-}$ mice. ACAT2$^{-/-}$ mice have nearly undetectable ACAT activity and lack cholesterol esters in the liver and small intestine. Because they cannot synthesize cholesterol esters in the small intestine, ACAT2$^{-/-}$ mice have a reduced capacity to obtain cholesterol from the diet. As a result, they do not develop hypercholesterolemia or form cholesterol gallstones when fed a diet high in fat, cholesterol, and cholic acid. ACAT2 deficiency markedly decreases the cholesterol ester content of apolipoprotein (apo) B-containing lipoproteins in mice fed either a chow diet or the above diet. Thus, ACAT2 is the cholesterol esterification enzyme in mouse liver and small intestine and plays a major role in mediating the response to dietary cholesterol.

Gene disruption studies suggest that both ACAT enzymes synthesize cholesterol esters: ACAT1 for storage in cytosolic droplets and ACAT2 for secretion in apoB-containing lipoproteins. However, these results may reflect their different tissue distributions rather than inherent biochemical differences. Both ACAT1 and ACAT2, when expressed in a cell line deficient in cholesterol esterification activity (AC29 cells), synthesize cholesterol esters that are stored in cytoplasmic droplets (25). Also, overexpression of ACAT1 in mouse liver causes both cholesterol ester storage and secretion.

ROLE OF ACAT1 IN ATHEROSCLEROSIS

Because cholesterol ester synthesis by ACAT1 is involved in macrophage foam cell formation, it has been hypothesized that inhibition of ACAT1 may reduce atherosclerosis. To test the hypothesis that ACAT1 deficiency in macrophages inhibits atherosclerosis, $ACAT1^{-/-}$ mice were crossed with two strains of mice that are susceptible to atherosclerosis: low density lipoprotein receptor-deficient ($LDLR^{-/-}$) mice and apoE-deficient ($apoE^{-/-}$) mice. ACAT1 deficiency did not prevent atherosclerotic lesions in either model. Because ACAT1 deficiency also reduced serum cholesterol levels in both strains of mice, its effect on lesion size was confounded by effects on serum cholesterol levels, and thus the contribution of ACAT1 to plaque formation could not be rigorously determined. Qualitative analysis of lesions in the mutant mice, however, revealed reduced amounts of cholesterol esters and macrophages. These alterations may increase lesion stability although this hypothesis remains to be tested.

Studies that analyzed the quantitative effect of ACAT1 deficiency on lesion development have produced different results. Ishibashi and colleagues crossed a different strain of $ACAT1^{-/-}$ mice with $LDLR^{-/-}$ and $apoE^{-/-}$ mice and found that ACAT1 deficiency was associated with ~50per cent reduction in aortic root lesions in $LDLR^{-/-}$ mice and did not decrease serum cholesterol levels. However, Fazio *et al.* showed that $LDLR^{/}$ mice reconstituted with $ACAT1^{-/-}$ macrophages had increased atherosclerotic lesions that contained more free cholesterol and fewer macrophages than control lesions. The apparent discrepancies between these studies may reflect differences in methodology or study design. Ishibashi and colleagues quantified lesion areas by measuring neutral lipid staining even though the cholesterol ester content of atherosclerotic lesions in $ACAT1^{-/-}$ mice was significantly diminished. The lesion areas consequently may have been underestimated. In the study by Fazio *et al*, selective inhibition of macrophage ACAT1 may not have reduced lesion size because ACAT activity was not inhibited in adjacent arterial smooth muscle cells, which can also become foam cells.

Both complete and macrophage-specific ACAT1 deficiencies have deleterious effects in hypercholesterolemic mice. ACAT1 deficiency causes extensive deposition of free cholesterol in the skin and brain of LDLR$^{-/-}$ and apoE$^{-/-}$ mice, suggesting an important role for the enzyme in cholesterol metabolism in these tissues when serum sterol levels are elevated. Although the dramatic effects in these models reflect the extreme condition of complete ACAT1 inhibition in the face of severe hypercholesterolemia, they indicate that selective ACAT1 inhibition can be detrimental.

FUNCTIONAL ANALYSIS OF DGAT

Because DGAT catalyzes the final step in the major pathway of triglyceride synthesis it was speculated that inactivation of the encoding gene in the mouse (*Dgat*) would cause severe morbidity or early lethality. Surprisingly, DGAT-deficient mice are viable and appear healthy although adult mice develop dry fur and female mice have a lactation defect. Even though these animals lack triglyceride synthesis in most tissues examined, *Dgat*$^{-/-}$ mice have normal serum triglyceride levels and triglycerides in white adipose tissue. These results indicate the existence of alternative pathways of triglyceride synthesis.

Although DGAT deficiency does not prevent triglyceride synthesis, *Dgat*$^{-/-}$ mice have~50per cent reduction in fat pad content and are resistant to weight gain when fed a high fat diet. Unexpectedly, these findings are associated with an ~15per cent increase in daily total energy expenditure, part of which is mediated by increased physical activity when the mice are fed a high fat diet. The mechanism for the increased energy expenditure is unknown, but the *Dgat*$^{-/-}$ phenotype indicates that the disruption of triglyceride synthesis significantly affects energy metabolism.

ACAT AND DGAT ENZYMES AS PHARMACEUTICAL TARGETS

Because ACAT enzymes participate in cholesterol ester synthesis for assembly of atherogenic apoB-containing lipoproteins and in macrophage foam cell formation, ACAT inhibitors have been developed as potential therapeutics. These agents produced mixed results in several animal models. In cholesterol-fed rabbits and hamsters, ACAT inhibitors reduced atherosclerotic lesion development independently of their effects on serum cholesterol levels. A study of genetic models of hypercholesterolemia in mice and rabbits suggested that ACAT inhibitors either had no effect or increased aortic sinus lesion size. One inhibitor (avasimibe) reduced atherosclerosis in several animal models and is being evaluated in clinical trials. This compound, like most currently available ACAT inhibitors, is not selective for one of the two ACAT enzymes. This pan-specificity may be desirable because pharmacologic and genetic studies in animals suggest that specific inhibition of macrophage ACAT1 may increase lesion size, possibly because of toxicity from excess free cholesterol. Selective inactivation of ACAT1 in hypercholesterolemia may also have detrimental consequences. Therefore, selective ACAT1 inhibition should be approached cautiously in humans. Partial inhibition of both ACAT enzymes may be a preferable strategy. A selective inhibitor of ACAT2 may be useful for preventing diet-induced hypercholesterolemia.

Recent findings in *Dgat*$^{-/-}$ mice (38) have sparked interest in DGAT inhibitors as possible therapeutic agents for obesity. Several naturally occurring compounds have been identified.

Signaling Via Hydrophobic Molecules

Hydrophobic molecules can move in and out of cells by passing through lipid bilayers. Nitric oxide, arachidonic acid and steroids have been shown to play important roles in cell signaling.

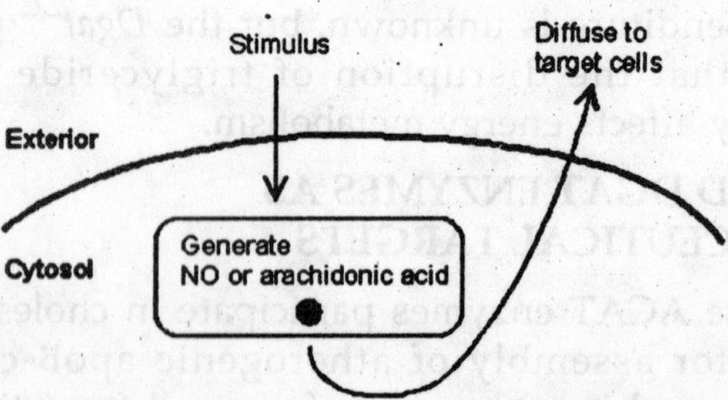

Fig. Signaling with NO or Arachidonic Acid. Unlike Most Signaling Cascades which Occur within the same cell, newly Generated NO or Arachidonic Acid Diffuses to Act on Target Molecules in Neighboring Cells.

NITRIC OXIDE

Nitric oxide (NO) is produced from the following reaction:

$$\text{L-arginine} \xrightarrow{\text{NOS}} \text{L-citrulline} + \text{NO}$$

Where NOS represents NO synthase. After NO is generated in a cell, it diffuses to act on target molecules in neighboring cells. NO may stimulate soluble guanylyl cyclase to produce cGMP which regulates several enzymes and ion channels. In smooth muscle, an important action of cGMP is to induce muscle relaxation. Normally, cGMP will soon be converted into GMP by phosphodiesterase. The well known drug for impotence, Viagra (sildenafil citrate), inhibits phosphodiesterase.

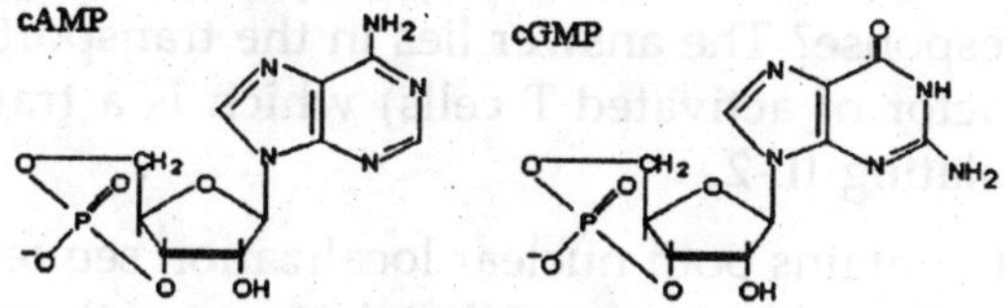

Fig. Structures of cAMP and cGMP

Arachidonic Acid

Arachidonic acid is generated from phospholipid hydrolysis catalyzed by phospholipase. Like NO, its target is usually not located in the cell where it was first generated. After diffusing to target cells, arachidonic acid may activate protein kinase, resulting in phosphorylation of target molecules. Many of its target molecules are involved in learning and other neuronal activities.

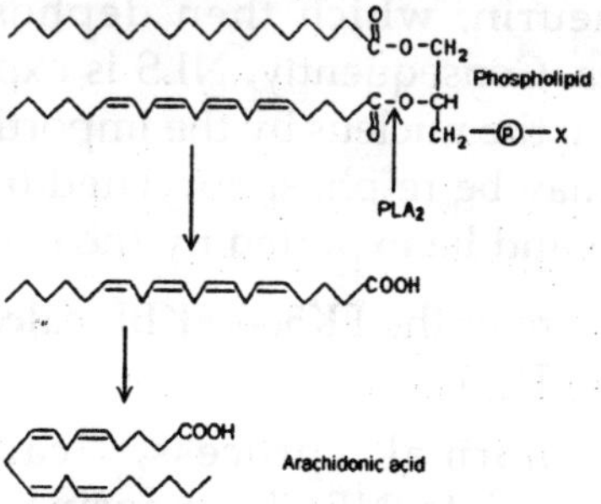

Fig. Generation of arachidonic acid from phospholipid by phospholipase A_2 (PLA_2).

Steroids

The major role of steroids is to regulate transcription, since many steroid receptors are transcription factors.

NFAT, Calcineurin and Immunosuppression

Organ transplantation often elicits immune responses to reject the foreign object. The most widely used drugs to deal with such graft reaction are FK506 and cyclosporin A (CsA). Their direct target is a phosphatase known as calcineurin. CsA combines with a cellular protein called cyclophilin to inhibit calcineurin; FK506 combines with FKBP (FK506 binding protein) to inhibit the phosphatase.

How could inhibition of a phosphatase suppress the immune response? The answer lies in the transport of NFAT (nuclear factor of activated T cells) which is a transcription factor regulating IL-2.

NFAT contains both nuclear localization sequence (NLS) and nuclear export sequence (NES), but one of them is buried in the protein interior, inaccessible to importin and exportin. Whether NLS or NES is masked depends on the phosphorylation state of specific serine residues in the regulatory domain. Phosphorylation of these serine residues exposes NES whereas dephosphorylation exposes NLS.

In resting cells, the NLS of the cytoplasmic NFAT is masked due to phosphorylation on these serine residues. In stimulated cells, an increase of intracellular calcium ions activates calcineurin, which then dephosphorylates the masking residues. Consequently, NLS is exposed and NFAT can be carried into the nucleus by the importin α/β. Inside the nucleus, NFAT may be re-phosphorylated by protein kinase, exposing its NES and be exported by the exportin Crm1.

a. The structure of the FK506-FKBP-calcineurin complex. PDB ID = 1TCO.

b. In the normal process, calcineurin may dephosphorylate NFAT, exposing its NLS so that NFAT can move into the nucleus to stimulate the expression of IL-2, inducing a cascade of immune responses. If calcineurin is inhibited, the immune response will be suppressed.

CLASSIFICATION OF ENZYMES

Based on catalyzed reactions, the nomenclature committee of the International Union of Biochemistry and Molecular Biology (IUBMB) recommended the following classification.

1. Oxidoreductases catalyze a variety of oxidation-reduction reactions. Common names include dehydrogenase, oxidase, reductase and catalase.
2. Transferases catalyze transfers of groups (acetyl, methyl, phosphate, etc.). Common names include acetyltransferase, methylase, protein kinase and polymerase. The first three subclasses play major roles in the regulation of cellular processes. The polymerase is essential for the synthesis of DNA and RNA.
3. Hydrolases catalyze hydrolysis reactions where a molecule is split into two or more smaller molecules by the addition of water. Common examples are given below.

Proteases splits protein molecules. Examples: HIV protease and caspase. HIV protease is essential for HIV replication. Caspase plays a major role in apoptosis.

Apoptosis

Apoptosis is a tightly regulated form of cell death, also called the programmed cell death. Morphologically, it is characterized by chromatin condensation and cell shrinkage in the early stage. Then the nucleus and cytoplasm fragment, forming membrane-bound apoptotic bodies which can be engulfed by phagocytes. In contrast, cells undergo another form of cell death, necrosis, swell and rupture. The released intracellular contents can damage surrounding cells and often cause inflammation.

Apoptosis is an important process during normal development. It is also involved in aging and various diseases such as cancer, AIDS, Alzheimer's disease and Parkinson's disease.

THE MECHANISM

During apoptosis, the cell is killed by a class of proteases called caspases. More than 10 caspases (web link) have been identified. Some of them (e.g., caspase 8 and 10) are involved in the initiation of apoptosis, others (caspase 3, 6, and 7) execute the death order by destroying essential proteins in the cell. The apoptotic process can be summarized as follows:

1. Activation of initiating caspases by specific signals.
2. Activation of executing caspases by the initiating caspases which can cleave inactive caspases at specific sites.
3. Degradation of essential cellular proteins by the executing caspases with their protease activity.
   - Death receptors: Fas/CD95, DR4/DR5, DR3, and TNFR (Tumor Necrosis Factor Receptor).
   - Adaptors: FADD (Fas-associated death domain protein) and TRADD (TNFR-associated death domain protein).
   - Activation: Binding of death ligands (FasL/CD95L, TRAIL/APO-2L, APO-3L and TNF) induces trimerization of their receptors, which then recruit adaptors and activate caspases.

     Note: TRADD is involved only in the coupling between caspases and DR3 or TNFR. This adaptor can also recruit other proteins (web link) to inhibit apoptosis through the NF-kB pathway.

As shown in the above figure, a variety of death ligands (FasL/CD95L, TRAIL, APO-3L and TNF) can induce apoptosis. It is natural to see if they can kill cancer cells without affecting normal cells. TNF was first investigated in the 1980s for cancer therapy, but with disappointing results. Then CD95L (FasL) was tested in the 1990s. The results were still not satisfactory. Recently, TRAIL has been demonstrated to be highly selective for transformed cells, with minimal effects on normal cells. It could be an effective drug for both cancer and AIDS.

Nucleases splits nucleic acids (DNA and RNA). Based on the substrate type, they are divided into RNase and DNase. RNase catalyzes the hydrolysis of RNA and DNase acts on DNA. They may also be divided into exonuclease and endonuclease. The exonuclease progressively splits off single nucleotides from one end of DNA or RNA. The endonuclease splits DNA or RNA at internal sites.

Phosphatase catalyzes dephosphorylation (removal of phosphate groups). Example: calcineurin.The immuno suppressive drugs FK506 and Cyclosporin A are the inhibitors of calcineurin.

NFAT, Calcineurin and Immunosuppression

Organ transplantation often elicits immune responses to reject the foreign object. The most widely used drugs to deal with such graft reaction are FK506 and cyclosporin A (CsA). Their direct target is a phosphatase known as calcineurin. CsA combines with a cellular protein called cyclophilin to inhibit calcineurin; FK506 combines with FKBP (FK506 binding protein) to inhibit the phosphatase.

How could inhibition of a phosphatase suppress the immune response? The answer lies in the transport of NFAT (nuclear factor of activated T cells) which is a transcription factor regulating IL-2.

NFAT contains both nuclear localization sequence (NLS) and nuclear export sequence (NES), but one of them is buried in the protein interior, inaccessible to importin and exportin. Whether NLS or NES is masked depends on the phosphorylation state of specific serine residues in the regulatory domain. Phosphorylation of these serine residues exposes NES whereas dephosphorylation exposes NLS.

In resting cells, the NLS of the cytoplasmic NFAT is masked due to phosphorylation on these serine residues. In stimulated cells, an increase of intracellular calcium ions activates calcineurin, which then dephosphorylates the masking residues. Consequently, NLS is exposed and NFAT can be carried into the nucleus by the importin a/b. Inside the nucleus, NFAT may be re-phosphorylated by protein kinase, exposing its NES and be exported by the exportin Crm1.

a. The structure of the FK506-FKBP-calcineurin complex. PDB ID = 1TCO.

b. In the normal process, calcineurin may dephosphorylate NFAT, exposing its NLS so that NFAT can move into the nucleus to stimulate the expression of IL-2, inducing a cascade of immune responses. If calcineurin is inhibited, the immune response will be suppressed.

4. Lyases catalyze the cleavage of C-C, C-O, C-S and C-N bonds by means other than hydrolysis or oxidation. Common names include decarboxylase and aldolase.
5. Isomerases catalyze atomic rearrangements within a molecule. Examples include rotamase, protein disulfide isomerase (PDI), epimerase and racemase.
6. Ligases catalyze the reaction which joins two molecules. Examples include peptide synthase, aminoacyl-tRNA synthetase, DNA ligase and RNA ligase.

The IUBMB committee also defines subclasses and sub-subclasses. Each enzyme is assigned an EC (Enzyme Commission) number. For example, the EC number of catalase is EC1.11.1.6. The first digit indicates that the enzyme belongs to oxidoreductase (class 1). Subsequent digits represent subclasses and sub-subclasses.

CATALYTIC MECHANISMS OF ENZYMES

From the energetic point of view, the reason why an enzyme can accelerate a reaction is because it can lower the energy barrier (activation energy) separating the substrate and the reaction product. For example, the covalent bond energy between two atoms ranges from 50 to 200 kcal/mol, which is far greater than the thermal energy (0.6 kcal/mol) at room temperature. Therefore, the covalent bond is unlikely to break in the absence of external interactions. Enzymes can provide a proper environment to lower the energy barrier.

The exact mechanism of lowering the energy barrier depends on individual systems. RNase A is a very interesting example. This enzyme can cleave an RNA molecule through hydrolysis reaction, but has no effect on DNA.

MEMBRANE PROTEINS

TRANSMEMBRANE PROTEINS

G-protein-Coupled Receptors

The major role of G-protein-coupled receptors is to transmit signals into the cell. They are characterized by seven transmembrane segments. This class of membrane proteins can respond to a wide range of agonists, including photon, amines, hormones, neurotransmitters and proteins. Some agonists bind to the extracellular loops of the receptor, others may penetrate into the transmembrane region.

Immunoglobulin Family

This protein family transmit signals for immune responses. Each of them contains one or two transmembrane segments. The immunoglobulin (i.e., antibody) consists of four peptide chains: two heavy chains and two light chains. V_H and C_H represent the variable and conserved domains of the heavy chain, respectively; V_L and C_L denote those for the light chain. Variable domains are the binding site of antigens. To send signals into the cell, this class of proteins are typically coupled to non-receptor protein tyrosine kinase located on the cytosolic face of the plasma membrane.

Transport Proteins

According to the Transport Commission (TC) system, all transport proteins may be divided into the following classes:

a. Channels/Pores

   Examples: ion channels, porins

b. Electrochemical Potential-driven transporters

   Examples: uniporters, symporters and antiporters

c. Primary Active Transporters

Examples: Na^+-K^+ ATPase, H^+ ATPase, ABC Transporter Superfamily and CFTR channel.

d. Group Translocators

e. Transport Electron Carriers

f. Auxiliary Transport Proteins

g. Incompletely Characterized

CHANNELS/PORES

Ion Channels [I]

Ion channels are a special class of proteins that conduct small ions such as Na^+, K^+, Ca^{2+} or Cl^-. The opening and closing of ion channels may depend on membrane voltage or the binding of molecules (the ligand). Its mechanism remains unclear. However, the mystery of their selectivity has been solved.

Ion Selectivity

Potassium Channels

By definition, a potassium channel conducts mainly K^+ ions while excluding Na^+ and Ca^{2+} ions. We note that a Na^+ ion is smaller than a K^+ ion. How could a channel conduct K^+ ions while excluding smaller ions? The answer lies in the ion's hydration energy.

The diameter of the potassium channel's selectivity filter (the narrowest region of its channel pore) should be about the same as the size of a Cs^+ ion (diameter = 3.3Å) because Cs^+ is the largest ion which can barely pass through the potassium channel. In a solution, ions are surrounded by water molecules. The first hydration shell of a K^+ or Na^+ ion contains six water molecules. The diameters of unhydrated Na^+ and K^+ ions are 1.96Å and 2.66Å, respectively. The effective diameter of a water molecule in a hydrated ion is larger than 2Å. Therefore, to pass through the selectivity filter, the K^+ or Na^+ ion must remove four water molecules from its first hydration shell, leaving only two (one at the front and anther at the back). The smaller Na^+ ion requires greater dehydration energy than the K^+ ion, because its nucleus has shorter distance with surrounding water molecules and thus interacting more strongly. As a result, the Na^+ ion is harder to pass through the potassium channel than the K^+ ion. The observation that lithium (Li^+) is excluded from the potassium channel can also be explained by the same energetic consideration. Divalent cations such as Ca^{2+} should also be excluded because their dehydration energies are much higher than monovalent cations.

Sodium Channels

In the sodium channel, the Na^+ ion is more permeable than the K^+ ion. This is because the selectivity filter of the sodium channel is slightly larger than that of the potassium channel. It is large enough to accommodate a Na^+ ion attached with three water molecules, but not enough for a K^+ ion attached with three water molecules. Therefore, to pass through the sodium channel, the Na^+ ion needs to remove only three, but the K^+ ion has to remove four, water molecules from its first hydration shell. The required dehydration energy for the K^+ ion is greater than the Na^+ ion.

Calcium Channels

In calcium channels, the permeability of monovalent cations (Na^+ and K^+) is about three orders of magnitude smaller than the Ca^{2+} permeability. This ion selectivity does not seem to involve hydration, because Ca^{2+} is more heavily hydrated than Na^+, and the unhydrated diameters of Ca^{2+} and Na^+ are almost identical. Then, how could calcium channels select Ca^{2+} over Na^+?

Although the permeability of monovalent cations in the calcium channel is quite small at normal ionic concentrations, large monovalent cationic current can be observed in the absence of Ca^{2+} and other divalent cations. This suggests that the calcium channel is basically permeable to both divalent and monovalent cations, but the selectivity arises from competition between ions. The calcium channel may contain a negatively charged binding site to facilitate ion conduction. The monovalent cations simply cannot compete with Ca^{2+} for this binding site. This idea has been demonstrated experimentally. In the calcium channel, if a negatively charged glutamate residue in the pore-lining region is mutated into a positively charged lysine, the calcium channel becomes more permeable to Na^+ than Ba^{2+}. Conversely, in the sodium channel, mutation of a pore-lining lysine residue into glutamate transforms the channel from a Na^+-selective to a Ca^{2+}-selective channel.

Channel Structures

The detailed 3D structures of most ion channels are not known. However, from the hydrophobicity profile of their amino acid sequences, it is possible to obtain the domain structures of most ion channels.

K Channels

There are many types of potassium channels. Most commonly observed K channels are composed of four subunits, each is homologous to the Shaker protein. The hydrophobicity profile indicates that it contains six hydrophobic segments, designated as S1 - S6. These segments are likely to be the transmembrane domains. Other experimental results suggests that the P-region is lining the channel pore.

Na and Ca Channels

A Na or Ca channel consists of a major pore-forming subunit and possibly other small auxiliary subunits. The major pore-forming subunit is called the α subunit which can be divided into four similar domains. Each domain is analogous to a Shaker protein with six transmembrane segments and a P-region. Thus, an α subunit is sufficient to form an ion channel.

Synaptic Channels

A large class of ion channels are specifically located at the synapses (the junction between nerve cells).

Porin

The nonpolar interior of membranes limits the free diffusion of polar molecules. In most membranes various proteins serve as channels and pumps which regulate the movement of ions, metabolites and even water across the membrane. Due to the specificity of these proteins there is a high degree of control over what passes through the membrane.Several membranes such as the outer membranes of mitochondria, chloroplasts and bacteria, however, do not need the same high degree of control over what gets across. Rather than synthesizing many selective proteins to carry polar molecules across, it appears that cells make only a few types of transporters, like the porin pictured below, which let almost anything below a certain size across. It may simply be more cost effective to put a few porins in these membranes then many more selective proteins. These "leaky" membranes are still effective barriers to the movement of large molecules such as proteins.

Below are two views of a model of porin from the bacterium *Rhodopseudomonas blastica.* In the left image the model is shown from the side as if you are in the middle of the membrane. The height of the protein is about the same as the thickness of the membrane in which it sits. The right image shows the model from the end where the central "hole" through which ions and metabolites pass is clearly visible. Regions of secondary structure are shown in yellow (beta sheet), pink (alpha helix) and blue (turn). The strands of colour follow the backbone of the molecule. If all of the atoms and bonds in the protein were visualized the image would be much too complicated.Why is the tertiary structure of porin described as a beta barrel?

Electrochemical Potential-driven Transporters

Uniporter-Catalyzed Transport

We begin our discussion of membrane transport proteins with the simplest type, which catalyze uniport transport. The plasma membrane of most cells contains several uniporters that enable amino acids, nucleosides, sugars, and other small molecules to enter and leave cells down their concentration gradients. Similar to enzymes, uniporters accelerate a reaction that is already thermodynamically favored, and the movement of a substance across a membrane down its concentration gradient will have the same negative ΔG value whether or not a protein transporter is involved. This type of movement sometimes is referred to as facilitated transport (or facilitated diffusion). Many chemical reactions that are thermodynamically favored will not occur unless an appropriate enzyme is present; such is also the case with movement of hydrophilic molecules across biological membranes. Unlike the substrates of enzymatic reactions, however, transported substances undergo no chemical change during movement across a membrane.

Three Main Features Distinguish Uniport Transport from Passive Diffusion

Three properties of uniporter-catalyzed movement of glucose and other small hydrophilic molecules across a membrane distinguish this type of transport from passive diffusion:

1. The rate of facilitated transport by uniporters is far higher than predicted by Fick's equation describing passive diffusion. Because the transported molecules never enter the hydrophobic core of the phospholipid bilayer, the partition coefficient K is irrelevant.
2. Transport is specific. Each uniporter transports only a single species of molecule or a single group of closely related molecules.
3. Transport occurs via a limited number of uniporter molecules, rather *than throughout the phospholipid bilayer*. Consequently, there is a maximum transport rate V_{max} that is achieved when the concentration gradient across the membrane is very large and each uniporter is working at its maximal rate.

Cotransport by Symporters and Antiporters

Besides ATP-powered pumps, cells have a second, discrete class of proteins that import or export ions and small molecules, such as glucose and amino acids, against a concentration gradient. These proteins use the energy stored in the electrochemical gradient of Na^+ or H^+ ions to power the uphill movement of another substance, a process called cotransport. For instance, the energetically favored movement of a Na^+ ion (the "cotransported" ion) into a cell across the plasma membrane, driven both by its concentration gradient and by the transmembrane voltage gradient can be coupled obligatorily to movement of the "transported" molecule (e.g., glucose) against its concentration gradient. When the transported molecule and cotransported ion move in the same direction, the process is called symport; when they move in opposite directions, the process is called antiport.

Primary Active Transporters

Active Transport by ATP-Powered Pumps

We turn now to the ATP-powered pumps that transport ions and various small molecules against their concentration gradients. The general structures of the four principal classes of these transport proteins are depicted in figure and their properties are summarized in table. Note that the P, F, and V classes transport ions only, whereas the ABC superfamily class transports small molecules as well as ions.

P-class ion pumps containα transmembrane catalytic á subunit, which contains an ATP-binding site, and usually a smaller β subunit, which may have regulatory functions. Many of these pumps are tetramers composed of twoα and two β subunits. During the transport process, at least one of the á subunits is phosphorylated (hence the label "P"), and the transported ions are thought to move through the phosphorylated subunit. This class includes the Na^+/K^+ ATPase in the plasma membrane, which maintains the Na^+ and K^+ gradients typical of animal cells, and several Ca^{2+} ATPases, which pump Ca^{2+} ions out of the cytosol into the external medium or into the lumen of the sarcoplasmic reticulum (SR) of muscle cells. Another member of the P class, found in acid-secreting cells of the mammalian stomach, transports protons (H^+ ions) out of and K^+ ions into the cell. The H^+ pump that maintains the membrane electric potential in plant, fungal, and bacterial cells also belongs to this class.

The structures of *F-class* and *V-class ion pumps* are similar to each other but unrelated to and more complicated than P-class pumps. F- and V-class pumps contain at least three kinds of transmembrane proteins and five kinds of extrinsic polypeptides that form the cytosolic domain. Several of the transmembrane and extrinsic subunits in F-class and V-class pumps exhibit sequence homology, and each pair of homologous subunits is thought to have evolved from a common polypeptide.

All known V and F pumps transport only protons in a process that does not involve a phosphoprotein intermediate. V-class pumps generally function to maintain the low pH of plant vacuoles and of lysosomes and other acidic vesicles in animal cells by using the energy released by ATP hydrolysis to pump protons from the cytosolic to the exoplasmic face of the membrane against the proton electrochemical gradient. F-class pumps are found in bacterial plasma membranes and in mitochondria and chloroplasts. In contrast to V pumps, they generally function to power the synthesis of ATP from ADP and P_i by movement of protons from the exoplasmic to the cytosolic face of the membrane down the proton electrochemical gradient.

The final class of ATP-powered transport proteins is larger and more diverse than the other classes. Referred to as the *ABC (ATP-binding cassette) superfamily,* this class includes more than 100 different transport proteins found in organisms ranging from bacteria to humans. Each ABC protein is specific for a single substrate or group of related substrates including ions, sugars, peptides, polysaccharides, and even proteins. All ABC transport proteins share a common organization consisting of four "core" domains: two transmembrane (T) domains, forming the passageway through which transported molecules cross the membrane, and two cytosolic ATP-binding (A) domains. In some ABC proteins, the core domains are present in four separate polypeptides; in others, the core domains are fused into one or two multidomain polypeptides.

All classes of ATP-powered pumps have one or more binding sites for ATP, and these are always on the cytosolic face of the membrane. Although these proteins are often called ATPases, they normally do not hydrolyze ATP into ADP and P_i unless ions or other molecules are simultaneously transported. Because of the tight coupling between ATP hydrolysis and transport, the energy stored in the phosphoanhydride bond is not dissipated. Thus ATP-powered transport proteins are able to collect the free energy released during ATP hydrolysis and use it to move ions or other molecules uphill against a potential or concentration gradient.

The energy expended by cells to maintain the concentration gradients of Na^+, K^+, H^+, and Ca^{2+} across the plasma and intracellular membranes is considerable. In nerve and kidney cells, for example, up to 25 percent of the ATP produced by the cell is used for ion transport; in human erythrocytes, up to 50 percent of the available ATP is used for this purpose. In cells treated with poisons that inhibit the aerobic production of ATP (e.g., 2,4-dinitrophenol), the ion concentration inside the cell gradually approaches that of the exterior environment as the ions move through plasma membrane channels down their electric and concentration gradients. Eventually treated cells die: partly because protein synthesis requires a high concentration of K^+ ions and partly because in the absence of a Na^+ gradient across the cell membrane, a cell cannot import certain nutrients such as amino acids. Studies on the effects of such poisons provided early evidence for the existence of ion pumps. In this section, we discuss in some detail examples of the P, V, and ABC classes of ATP-powered pumps.

Plasma-Membrane Ca^{2+} ATPase Exports Ca^{2+} Ions from Cells.

In order for Ca^{2+} to function in intracellular signaling, its cytosolic concentration usually must be kept below 0.1 – 0.2 μM. (Although some cytosolic Ca^{2+} is bound to negatively charged groups, it is the concentration of *free,* unbound Ca^{2+} that is critical to its signaling function.) The plasma membranes of animal, yeast, and probably plant cells contain Ca^{2+} ATPases that transport Ca^{2+} out of the cell against its electrochemical gradient. These P-class ion pumps help maintain the concentration of free Ca^{2+} ions in the cytosol at a low level.

In addition to a catalytica subunit containing an ATP-binding site, as found in other P-class pumps, plasmamembrane Ca^{2+} ATPases also contain the Ca^{2+}-binding regulatory protein calmodulin. A rise in cytosolic Ca^{2+} induces the binding of Ca^{2+} ions to calmodulin, which triggers an allosteric activation of the Ca^{2+} ATPase; as a result, the export of Ca^{2+} ions from the cell accelerates, and the original low cytosolic concentration of free Ca^{2+} is restored rapidly.

Muscle Ca^{2+} ATPase Pumps Ca^{2+} Ions from the Cytosol into the Sarcoplasmic Reticulum.

Besides the plasma-membrane Ca^{2+} ATPase, muscle cells contain a second, different Ca^{2+} ATPase that transports Ca^{2+} from the cytosol into the lumen of the sarcoplasmic reticulum (SR), an internal organelle that concentrates and stores Ca^{2+} ions. As discussed the SR and its calcium pump (referred to as the *muscle calcium pump*) are critical in muscle contraction and relaxation: release of Ca^{2+} ions from the SR into the muscle cytosol causes contraction, and the rapid removal of Ca^{2+} ions from the cytosol by the muscle calcium pump induces relaxation.

Because the muscle calcium pump constitutes more than 80 percent of the integral protein in SR membranes, it is easily purified and characterized. Each transmembrane catalytica subunit has a molecular weight of 100,000 and transports two Ca^{2+} ions per ATP hydrolyzed. In the cytosol of muscle cells, the free Ca^{2+} concentration ranges from 10^{-7} M (resting cells) to more than 10^{-6} M (contracting cells), whereas the *total* Ca^{2+} concentration in the SR lumen can be as high as 10^{-2} M. Sites on the cytosolic surface of the muscle calcium pump have a very high affinity for Ca^{2+} ($K_m = 10^{-7}$ M), allowing the pump to transport Ca^{2+} efficiently from the cytosol into the SR against the steep concentration gradient.

The concentration of free Ca^{2+} within the sarcoplasmic reticulum is actually much less than the total concentration of

10^{-2} M. Two soluble proteins in the lumen of SR vesicles bind Ca^{2+} and serve as a reservoir for intracellular Ca^{2+}, thereby reducing the concentration of free Ca^{2+} ions in the SR vesicles, and consequently decreasing the energy needed to pump Ca^{2+} ions into them from the cytosol. The activity of the muscle Ca^{2+} ATPase is so regulated that if the free Ca^{2+} concentration in the cytosol becomes too high, the rate of calcium pumping increases until the cytosolic Ca^{2+} concentration is reduced to less than 1 μM. Thus in muscle cells, the calcium pump in the SR membrane can supplement the activity of the plasma-membrane pump, assuring that the cytosolic concentration of free Ca^{2+} remains below 1 μM.

The current model of the mechanism of action of the Ca^{2+} ATPase in the SR membrane is outlined in Figure. Coupling of ATP hydrolysis with ion pumping involves several steps that must occur in a defined order.When the protein is in one conformation, termed *E1*, two Ca^{2+} ions bind in sequence to high-affinity sites on the cytosolic surface (step 1). Then an ATP binds to its site on the cytosolic surface; in a reaction requiring that a Mg^{2+} ion be tightly complexed to the ATP, the bound ATP is hydrolyzed to ADP and the liberated phosphate is transferred to a specific aspartate residue in the protein, forming a high-energy acyl phosphate bond, denoted by E1~P (step 2). The protein then changes its conformation to E2 – P, generating two lowaffinity Ca^{2+}-binding sites on the exoplasmic surface, which faces the SR lumen; this conformational change simultaneously propels the two Ca^{2+} ions through the protein to these sites (step 3) and inactivates the high-affinity Ca^{2+}-binding sites on the cytosolic face. The Ca^{2+} ions then dissociate from the exoplasmic surface of the protein (step 4). Following this, the aspartyl-phosphate bond in E2 – P is hydrolyzed, causing E2 to revert to E1, a change that inactivates the exoplasmic-facing Ca^{2+}-binding sites and regenerates the cytosolicfacing Ca^{2+}-binding sites (step 5).

Thus phosphorylation of the muscle calcium pump by ATP favors conversion of E1 to E2, and dephosphorylation favors the conversion of E2 to E1. While only E2 – P, not E1~P, is actually hydrolyzed, the free energy of hydrolysis of the aspartyl-phosphate bond in E1~P is greater than that for E2 – P. The reduction in free energy of the aspartyl-phosphate bond in E2 – P, relative to E1~P, can be said to power the E1 E2 conformational change. The affinity of Ca^{2+} for the cytosolic-facing binding sites in E1 is a thousandfold greater than the affinity of Ca^{2+} for the exoplasmic-facing sites in E2; this difference enables the protein to transport Ca^{2+} unidirectionally from the cytosol, where it binds tightly to the pump, to the exoplasm, where it is released.

For instance, the muscle calcium pump has been isolated with phosphate linked to an aspartate residue, and spectroscopic studies have detected slight alterations in protein conformation during the E1 E2 conversion. On the basis of the protein's amino acid sequence and various biochemical studies, investigators proposed the structural model for the catalyticα subunit shown in Figure. The membrane-spanningα helices are thought to form the passageway through which Ca^{2+} ions move. The bulk of the subunit consists of cytosolic globular domains that are involved in ATP binding, phosphorylation of aspartate, and energy transduction. These domains are connected by "stalks" to the membrane-embedded domain.

As noted previously, all P-class ion pumps, regardless of which ion they transport, are phosphorylated during the transport process. The amino acid sequences around the phosphorylated aspartate in the catalyticα subunit are highly conserved in all proteins of this type. Thus the mechanistic model in Figure probably is generally applicable to all these ATP-powered ion pumps. In addition, theα subunits of all the P pumps examined to date have a similar molecular weight and, as deduced from their amino acid sequences derived from cDNA clones, have a similar arrangement of transmembraneα helices. These findings strongly suggest that all these proteins evolved from a common precursor, although they now transport different ions.

Na^{+}/K^{+} ATPase Maintains the Intracellular Na^{+} and K^{+} Concentrations in Animal Cells.

A second P-class ion pump that has been studied in considerable detail is the Na^+/K^+ ATPase present in the plasma membrane of all animal cells. This ion pump is a tetramer of subunit composition $\alpha_2\beta_2$. (Classic Experiment 15.1 describes the discovery of this enzyme.) The β polypeptide is required for newly synthesizedα subunits to fold properly in the endoplasmic reticulum but apparently is not involved directly in ion pumping.

Theα subunit is a 120,000-MW nonglycosylated polypeptide whose amino acid sequence and predicted membrane structure are very similar to those of the muscle Ca^{2+} ATPase. In particular, the Na^+/K^+ ATPase has a stalk on the cytosolic face that links domains containing the ATP-binding site and the phosphorylated aspartate to the membrane-embedded domain. The overall process of transport moves three Na^+ ions out of and two K^+ ions into the cell per ATP molecule split.

Several lines of evidence indicate that the Na^+/K^+ ATPase is responsible for the coupled movement of K^+ and Na^+ into and out of the cell, respectively. For example, the drug *ouabain,* which binds to a specific region on the exoplasmic surface of the protein and specifically inhibits its ATPase activity, also prevents cells from maintaining their Na^+/K^+ balance. Any doubt that the Na^+/K^+ ATPase is responsible for ion movement was dispelled by the demonstration that the enzyme, when purified from the membrane and inserted into liposomes, propels K^+ and Na^+ transport in the presence of ATP.

The mechanism of action of the Na^+/K^+ ATPase, outlined in figure is similar to that of the muscle calcium pump, except that ions are pumped in both directions across the membrane. In its E1 conformation, the Na^+/K^+ ATPase has three high-affinity Na^+-binding sites and two low-affinity K^+-binding sites on the cytosolic-facing surface of the protein. The K_m for binding of Na^+ to these cytosolic sites is 0.6 mM, a value considerably lower than the intracellular Na^+ concentration of H"12 mM; as a result, Na^+ ions normally will fill these sites. Conversely, the affinity of the cytosolic K^+-binding sites is low enough that K^+ ions, transported inward through the protein, dissociate from E1 into the cytosol despite the high intracellular K^+ concentration. During the E1 E2 transition, the three bound Na^+ ions move outward through the protein.

Transition to the E2 conformation also generates two high-affinity K^+ sites and three low-affinity Na^+ sites on the exoplasmic face. Because the K_m for K^+ binding to these sites (0.2 mM) is considerably lower than the extracellular K^+ concentration (4 mM), these sites will fill quickly with K^+ ions. In contrast, the three Na^+ ions, transported outward through the protein, will dissociate into the extracellular medium from the low-affinity Na^+ sites on the exoplasmic surface despite the high extracellular Na^+ concentration. Similarly, during the E2 E1 transition, the two bound K^+ ions are transported inward.

V-Class H^+ ATPases Pump Protons across Lysosomal and Vacuolar MembranesAll V-class ATPases transport H^+ ions only. These proton pumps, present in the membranes of lysosomes, endosomes, and plant vacuoles, function to acidify the lumen of these organelles. The acidity of the lysosomal lumen, usually H"4.5 – 5.0, can be measured precisely in living cells by use of particles labeled with a pH-sensitive fluorescent dye. Cells phagocytose these particles and transfer them to the lysosomes. The ability of different wavelengths of visible light to excite fluorescence is highly dependent on pH, and the lysosomal pH can be calculated from the spectrum of the fluorescence emitted. Maintenance of the 100-fold or more proton gradient between the lysosomal lumen (pH H"4.5 – 5.0) and the cytosol (pH H"7.0) depends on ATP production by the cell.

The ATP-powered proton pumps in lysosomal and vacuolar membranes have been isolated, purified, and incorporated into liposomes. As illustrated in figure these V-class proton pumps contain two discrete domains: a cytosolic-facing hydrophilic domain (V_1) composed of five different polypeptides and a transmembrane domain (V_0) containing 9 – 12 copies of proteolipid c, one copy of protein b, and one copy of protein a. The subunit composition of the cytosolic domain is $\alpha_3\beta_3\gamma\delta\varepsilon$; the α and β subunits contain the sites where ATP binding and hydrolysis occur. Each transmembrane c subunit is thought to span the membrane two times; the c and a subunits together form the proton-conducting channel. Unlike P-class ion pumps, the V-class H^+ ATPases are not phosphorylated and dephosphorylated during proton transport.

Similar V-class ATPases are found in the plasma membrane of certain acid-secreting cells. These include osteoclasts, bone-resorbing macrophagelike cells, which bind to a bone and seal off a small segment of extracellular space between the plasma membrane and the surface of the bone. HCl secreted into this space by osteoclasts dissolves the calcium phosphate crystals that give bone its rigidity and strength.

Another example is the mitochondria-rich epithelial cells lining the toad bladder; the apical plasma membrane of these cells contain many V-class H^+ ATPases, which function to acidify the urine. As we discuss later, the membrane of plant vacuoles contains two proton pumps: a typical V-class H^+ ATPase and another one that utilizes the energy released by hydrolysis of inorganic pyrophosphate (PP_i) to pump protons into the vacuole. This PP_i-hydrolyzing proton pump, believed to be unique to plants, has an amino acid sequence different from any other ion-transporting proteins

ATP-powered proton pumps cannot acidify the lumen of an organelle (or the extracellular space) by themselves. The reason for this is that pumping of protons would rapidly cause a buildup of positive charge on the exoplasmic face of the membrane on the inside of the vesicle membrane and a corresponding buildup of negative charges on the cytosolic face. In other words, the pump would generate a voltage across the membrane, exoplasmic face positive, which would prevent movement of protons into the vesicle before a significant H^+ concentration gradient had been established. In fact, this is the way that H^+ pumps generate an insidenegative potential across plant and yeast plasma membranes. In order for an organelle lumen or an extracellular space (e.g., the outside of an osteoclast) to become acidic, movement of H^+ up its concentration gradient must be accompanied by

1. Movement of an equal number of anions in the same direction or
2. Movement of equal numbers of a different cation in the opposite direction. The first process occurs in lysosomes and plant vacuoles whose membranes contain V-class H^+ ATPases and ion channels through which accompanying anions (e.g., Cl″) move. The second occurs in the lining of the stomach, which contains a P-class H^+/K^+ ATPase that pumps one H^+ outward and one K^+ inward.

The ABC Superfamily Transports a Wide Variety of Substrates

As noted earlier, all members of the very large and diverse ABC superfamily of transport proteins contain two transmembrane (T) domains and two cytosolic ATP-binding (A) domains. The T domains, each built of six membrane-spanning á helices, form the pathway through which the transported substance (substrate) crosses the membrane and determine the substrate specificity of each ABC protein. The sequence of the A domains is H″30 to 40 percent homologous in all members of this superfamily, indicating a common evolutionary origin. Some ABC proteins also contain a substrate-binding subunit or regulatory subunit.

Bacterial Plasma-Membrane Permeases

The plasma membrane of many bacteria contain numerous *permeases* that belong to the ABC superfamily. These proteins use the energy released by hydrolysis of ATP to transport specific amino acids, sugars, vitamins, or even peptides into the cell. Since bacteria frequently grow in soil or pond water where the concentration of nutrients is low, these ABC transport proteins allow the cells to concentrate amino acids and other nutrients in the cell against a substantial concentration gradient. Bacterial permeases generally are *inducible;* that is, the quantity of a transport protein in the cell membrane is regulated by both the concentration of the nutrient in the medium and the metabolic needs of the cell.

In *E. coli* histidine permease, a typical bacterial ABC protein, the two transmembrane domains and two cytosolic ATP-binding domains are formed by four separate subunits. In gram-negative bacteria such as *E. coli,* which have an outer membrane, a soluble histidine-binding protein in the periplasmic space assists in transport. This soluble protein binds histidine tightly and directs it to the T subunits, through which histidine crosses the membrane powered by ATP hydrolysis. Mutant *E. coli* cells that are defective in any of the histidine-permease subunits or the soluble binding protein are unable to transport histidine into the cell, but are able to transport other amino acids whose uptake is facilitated by other transport proteins. Such genetic analyses provide strong evidence that histidine permease and similar ABC proteins function to transport solutes into the cell.

Mammalian MDR Transport Proteins

A series of rather unexpected observations led to discovery of the first eukaryotic ABC protein. Oncologists noted that tumor cells often became simultaneously resistant to several chemotherapeutic drugs with unrelated chemical structures; similarly, cell biologists observed that cultured cells selected for resistance to one toxic substance (e.g., colchicine, a microtubule inhibitor) frequently became resistant to several other drugs, including the anticancer drug adriamycin. Subsequent studies showed that this resistance is due to enhanced expression of a*multidrug-resistance (MDR) transport protein* known as *MDR1.* In this member of the ABC superfamily, all four domains are "fused" into a single 170,000-MW protein. This protein uses the energy derived from ATP hydrolysis to*export* a large variety of drugs from the cytosol to the extracellular medium. The *Mdr1* gene is frequently amplified in multidrug-resistant cells, resulting in a large overproduction of the MDR1 protein.

Most drugs transported by MDR1 are small hydrophobic molecules, which diffuse from the culture medium across the plasma membrane into the cell. The ATP-powered export of such drugs from the cytosol by MDR1 means a much higher extracellular drug concentration is required to kill cells. That MDR1 is an ATP-powered small-molecule pump has been demonstrated with liposomes containing the purified protein. The ATPase activity of these liposomes is enhanced by different drugs in a dose-dependent manner corresponding to their ability to be transported by MDR1.

Not only does MDR1 transport a varied group of molecules, but all these substrates compete with one another for transport by MDR1. Although the mechanism of action of MDR1-assisted transport has not been definitively demonstrated, the *flippase model*, is a likely candidate. Substrates of MDR1 are primarily planar, lipid-soluble molecules with one or more positive charges, and they move spontaneously from the cytosol into the cytosolic-facing leaflet of the plasma membrane. The hydrophobic portion of a substrate molecule is oriented toward the hydrophobic core of the membrane, and the charged portion toward the polar cytosolic face of the membrane and is still in the cytosol. The substrate diffuses laterally until encountering and binding to a site on the MDR1 protein that is within the bilayer. The protein then "flips" the charged substrate molecule into the exoplasmic leaflet, an energetically unfavorable reaction powered by the coupled ATPase activity of MDR1.

Once in the exoplasmic face, the substrate diffuses into the aqueous phase on the outside of the cell. Support for the flippase model of transport by MDR1 comes from MDR2, a homologous protein present in the region of the liver cell plasma membrane that faces the bile duct. MDR2 has been shown to flip phospholipids from the cytosolic-facing leaflet of the plasma membrane to the exoplasmic leaflet, thereby generating an excess of phospholipids in the exoplasmic leaflet; these phospholipids peel off into the bile duct and form an essential part of the bile. An alternative *pump model* also has been proposed for MDR1. According to this model, drug molecules in the cytosol bind directly to a single small-molecule binding site on the cytosolic face of the MDR1 protein; subsequent ATP hydrolysis powers movement of the bound drug through the protein to the aqueous phase on the outside of the cell by a mechanism similar to that of other ATP-powered pumps.

MDR1 protein is expressed in abundance in the liver, intestines, and kidney — sites from which natural toxic products are removed from the body. Thus the natural function of MDR1 may be to transport a variety of natural and metabolic toxins into the bile, intestinal lumen, or forming urine. During the course of its evolution, MDR1 appears to have coincidentally acquired the ability to transport drugs whose structures are similar to those of these toxins. Tumors derived from these cell types, such as hepatomas (liver cancers), frequently are resistant to virtually all chemotherapeutic agents and thus difficult to treat, presumably because the tumors exhibit increased expression of the MDR1 or MDR2 proteins.

Cystic Fibrosis Transmembrane Regulator (CFTR) Protein

Discovery of another ABC transport protein came from studies of cystic fibrosis (CF), the most common lethal autosomal recessive genetic disease of Caucasians. This disease is caused by a mutation in the *CFTR* gene, which encodes a chloride-channel protein that is regulated by cyclic AMP (cAMP), an intracellular second messenger. These Cl″ channels are present in the apical plasma membranes of epithelial cells in the lung, sweat glands, pancreas, and other tissues. An increase in cAMP stimulates Cl″ transport by such cells from normal individuals, but not from CF individuals who have a defective CFTR protein.

The sequence and predicted structure of the encoded CFTR protein, based on analysis of the cloned gene, are very similar to those of MDR1 protein except for the presence of an additional domain, the regulatory (R) domain, on the cytosolic face. The Cl″-channel activity of CFTR protein clearly is enhanced by binding of ATP. Moreover, cAMP activates a protein kinase that phosphorylates, and thereby activates, CFTR. When purified CFTR protein is incorporated into liposomes, it forms Cl″ channels with properties similar to those in normal epithelial cells. And when the wild-type CFTR protein is expressed by recombinant techniques in cultured epithelial cells from CF patients, the cells recover normal Cl″-channel activity. This latter result raises the possibility that gene therapy might reverse the course of cystic fibrosis.

Since CFTR protein is similar to MDR1 in structure, it may also function as an ATP-powered pump of some as-yet unidentified molecule. In any case, much remains to be learned about this fascinating class of ABC transport proteins.

MEMBRANE-ATTACHED PROTEINS

Membrane-attached proteins can be divided into three groups:

- Bound to the cytosolic face of the plasma membrane by myristate;
- Bound to the cytosolic face of the plasma membrane by farnesyl;
- Bound to the extracelluar face of the plasma membrane by glycosylphosphatidylinositol.

MISCELLANEOUS PROTEINS

In addition to enzymes and membrane proteins, there are many other proteins with various cellular functions. This page is not intended to be comprehensive, but lists only a few important examples.

CHAPERONES

Involved in protein folding.

The Small Heat Shock Proteins and Their Role in Human Disease

Small heat shock proteins (sHSPs) function as molecular chaperones, preventing stress induced aggregation of partially denatured proteins and promoting their return to native conformations when favorable conditions pertain. Sequence similarity between sHSPs resides predominately in an internal stretch of residues termed the crystallin domain, a region usually flanked by two extensions. The poorly conserved N-terminal extension influences oligomer construction and chaperone activity, whereas the flexible C-terminal extension stabilizes quaternary structure and enhances protein/substrate complex solubility. sHSP polypeptides assemble into dynamic oligomers which undergo subunit exchange and they bind a wide range of cellular substrates. As molecular chaperones, the sHSPs protect protein structure and activity, thereby preventing disease, but they may contribute to cell malfunction when perturbed. For example, sHSPs prevent cataract in the mammalian lens and guard against ischemic and reperfusion injury due to heart attack and stroke. On the other hand, mutated sHSPs are implicated in diseases such as desmin-related myopathy and they have an uncertain relationship to neurological disorders including Parkinson's and Alzheimer's disease. This review explores the involvement of sHSPs in disease and their potential for therapeutic intervention.

Within the molecular chaperone family, sHSPs constitute a structurally divergent group characterized by a conserved sequence of 80–100 amino acid residues termed the -crystallin domain. The -crystallin domain, duplicated in the unusual example of parasitic flatworms (Platyhelminthes), is located toward a highly flexible, variable, C-terminal extension, and is usually preceded by a poorly conserved N-terminal region. The molecular mass of sHSP subunits ranges from 12 to 43 kDa, and they assemble into large, dynamic complexes up to 1 MDa. sHSP secondary structure is dominated by -strands with limited -helical content, and -sheets within the -crystallin domain mediate dimer formation. Crystallization of two sHSPs has contributed significantly to the description of oligomerization, quaternary structure, subunit exchange, and chaperone activity. Characterization of a highly conserved arginine is also an important outcome of crystallization and related studies because mutation of this residue has profound effects on sHSP function and contributes to certain diseases.

The sHSPs are molecular chaperones, storing aggregation prone proteins as folding competent intermediates and conferring enhanced stress resistance on cells by suppressing aggregation of denaturing proteins, actions associated with oligomerization and subunit exchange. Functional studies of the sHSPs are more limited than for other chaperones, but this is changing as the application of genomics and proteomics reveals sHSP characteristics and their medical importance emerges. In this context, 10 sHSPs, termed HspB1–10, many of which are constitutively present at high levels in muscle and implicated in disease, are found in humans. Intracellular quantities and cellular localizations of sHSPs change in response to development, physiological stressors such as anoxia/hypoxia, heat and oxidation, and in relation to pathological status. sHSPs interact with many essential cell structures and it follows from such promiscuity that functional disruption and inappropriate association of these molecular chaperones with substrates will foster disease. Therefore, this review considers the role of sHSPs in several human medical conditions and it ends with a discussion of their therapeutic potential.

sHSPs were described previously as the 'forgotten chaperones', but this is no longer true. Two sHSPs have been crystallized, opening the door to more informed interpretation of results obtained by site-directed mutagenesis and other molecular probing. The functions of sHSP domains and individual amino acid residues are becoming clearer, as is the molecular basis of oligomerization. The implications of oligomer assembly and disassembly as chaperoning prerequisites are under study, sHSP substrates have been identified, and the role of ATP-dependent chaperones in substrate release and refolding revealed. sHSPs operate in the front lines of cell defence, protecting proteins during stress and providing opportunities for salvage. As molecular chaperones, sHSPs have the potential to guard cells from disease, but when perturbed or as residents of aberrant cells, they may promote disease. For example, sHSPs defend against ischemia, oxidative damage and apoptosis, but post-translational modifications and gene mutations cause cataract and desmin-related myopathies. Disease involvement suggests therapeutic exploitation of sHSPs, but this remains poorly explored, as is generally true for other HSPs. However, as sHSPs are better understood, opportunities for disease prevention and treatment become more apparent, and this, along with their fundamental importance in stress physiology, means that sHSPs will not be forgotten for some time to come.

The Role of Structural Disorder in the Function of RNA and Protein Chaperones

Chaperones are highly sophisticated protein machines that assist the folding of RNA molecules or other proteins. Their function is generally thought to require a fine-tuned and highly conserved structure: despite the recent recognition of the widespread occurrence of structural disorder in the proteome, this structural trait has never been generally considered in molecular chaperones. In this review we give evidence for the prevalence of functional regions without a well-defined 3-D structure in RNA and protein chaperones. By considering a variety of individual examples, we suggest that the structurally disordered chaperone regions either function as molecular recognition elements that act as solubilizers or locally loosen the structure of the kinetically trapped folding intermediate via transient binding to facilitate its conformational search.

The importance of structural disorder is underlined by a predictor of natural disordered regions, which shows an extremely high proportion of such regions, unparalleled in any other protein class, within RNA chaperones: 54.2per cent of their residues fall into disordered regions and 40per cent fall within disordered regions longer than 30 consecutive residues. Structural disorder also prevails in protein chaperones, for which frequency values are 36.7per cent and 15per cent, respectively. In keeping with these and other details, a novel "entropy transfer" model is presented to account for the mechanistic role of structural disorder in chaperone function.— Tompa, P., Csermely, P. The role of structural disorder in the function of RNA and protein chaperones.

Intrinsically disordered/unstructured proteins and protein domains exist and function without a well-defined 3-dimensional fold. Although these proteins defy the classical structure-function paradigm, they are not exceptions that violate the rule: they are rather common in living organisms and fulfill essential cellular functions. The prevalence of intrinsic disorder has been determined in different functional classes of proteins, and the highest level was found in regulatory, cell signaling, and cancer-associated proteins. Disorder and flexibility have never been comprehensively assessed in chaperones, although sporadic data suggest their presence and functional importance in this protein class.

Chaperones assist RNA molecules or other proteins in reaching their functional 3-dimensional structure. Often, folding of long polypeptide or polynucleotide chains is impeded by kinetic traps (i.e., local energy minima), which are stable enough to halt the folding process for a physiologically significant amount of time. Furthermore, such partially folded or misfolded intermediates tend to undergo irreversible aggregation, which poses a particular danger to the organism. This folding problem can be overcome by the aid of specific chaperone proteins that prevent aggregation and unfold the intermediate to offer another chance for a folding attempt. Chaperone function is structurally and mechanistically rather demanding, as chaperones have to assist the folding of a wide range of unrelated RNAs/proteins in the crowded intracellular milieu. No wonder the appearance of such highly sophisticated protein machines was a critical early evolutionary invention.

The intriguing aspect of disorder in chaperone function and evolution is highlighted by a recent suggestion that the first proteins to come into being were RNA chaperones. It is generally accepted that the biosphere once was dominated by organisms in which RNA was used for information storage and catalysis. Because RNA is especially prone to misfolding, simple proteins with chaperone activity must have conferred a significant selective advantage in a world so highly dependent on RNA function. As the first proteins must have been short and unfolded polypeptides, the presence of disordered sequences in protein chaperones and the recent discovery of fully disordered chaperone-like proteins and structural disorder in RNA chaperones underline the possibility of the critical early contribution, and the likely contemporary significance, of structural disorder in chaperone function.

In this chapter, the experimental evidence for the occurrence and functional role of structural disorder in RNA and protein chaperones is reviewed. We show that disordered segments occur in a wide range of chaperones and such regions are often indispensable in functional terms. The general occurrence of structural disorder is further demonstrated by the newly developed predictor for protein disorder (PONDR) algorithm: the frequency of disorganized structure is higher among RNA chaperones than in any other protein class, and is also very high for protein chaperones. The model of entropy transfer for disorder in chaperone function is suggested to contain all the individual examples and the very high frequency of structural disorder observed.

Chaperone-Mediated Protein Folding

The folding of most newly synthesized proteins in the cell requires the interaction of a variety of protein cofactors known as molecular chaperones. These molecules recognize and bind to nascent polypeptide chains and partially folded intermediates of proteins, preventing their aggregation and misfolding. There are several families of chaperones; those most involved in protein folding are the 40-kDa heat shock protein (HSP40; DnaJ), 60-kDa heat shock protein (HSP60; GroEL), and 70-kDa heat shock protein (HSP70; DnaK) families. The availability of high-resolution structures has facilitated a more detailed understanding of the complex chaperone machinery and mechanisms, including the ATP-dependent reaction cycles of the GroEL and HSP70 chaperones. For both of these chaperones, the binding of ATP triggers a critical conformational change leading to release of the bound substrate protein. Whereas the main role of the HSP70/HSP40 chaperone system is to minimize aggregation of newly synthesized proteins, the HSP60 chaperones also facilitate the actual folding process by providing a secluded environment for individual folding molecules and may also promote the unfolding and refolding of misfolded intermediates.

The basic paradigm of molecular chaperones is that they recognize and selectively bind nonnative, but not native, proteins to form relatively stable complexes. In most cases, the complexes are dissociated by the binding and hydrolysis of ATP. In addition, there are "specific" molecular chaperones that typically are involved in the assembly of particular multiprotein complexes. Molecular chaperones comprise several highly conserved families of unrelated proteins; many chaperones are also heat shock (stress) proteins. The ubiquitous role of molecular chaperones continues to unfold with more discoveries each year. In the context of in vivo protein folding, chaperones prevent irreversible aggregation of nonnative conformations and keep proteins on the productive folding pathway. In addition, they may maintain newly synthesized proteins in an unfolded conformation suitable for translocation across membranes and bind to nonnative proteins during cellular stress, among other functions. It is likely that most, if not all, cellular proteins will interact with a chaperone at some stage of their lifetime.

The focus of this review is on the functional contribution of chaperones to in vivo protein folding and assembly, especially those chaperones that are promiscuous, in that they show broad specificity for binding nonnative proteins This review is of necessity selective; the main goal is to furnish an up-to-date overview of the role of the major molecular chaperones involved in protein folding. In view of the vast array of literature on molecular chaperones, and the ease of access to literature citations using the Internet, this article should not be viewed as exhaustive.

Why do we need chaperones? After all, a basic tenet of in vitro protein folding has been the seminal work of Anfinsen which demonstrated that formation of the native protein from the unfolded state is a spontaneous process determined by the global free energy minimum. The results indicated that the native state of small globular proteins is determined by their amino acid sequence. However, the experimental conditions necessary to successfully fold many proteins, especially larger ones, in vitro, are very constrictive, usually requiring very low protein concentration and long incubation times and are usually unphysiological (e.g., relatively low temperatures). In contrast, most cells operate at ambient or homeothermically set temperatures (e.g., 37°C) where the hydrophobic effect will be stronger and thus protein denaturation and aggregation will be bigger problems, and the time-frame available for successful folding is short. Thus there is the need for additional factors for the successful folding of many proteins in vivo. When one considers the crowded cellular environment within a cell, it becomes clear that in vitro folding experiments at low protein concentrations are poor models for what happens in the cell, where a newly synthesized protein is in an environment with little or no "free" water, very high concentrations of other proteins and metabolites, and typically membranes, cytoskeletal elements, and other cellular components. Thus the need for chaperones

1. To prevent aggregation and misfolding during the folding of newly synthesized chains,
2. To prevent nonproductive interactions with other cell components,
3. To direct the assembly of larger proteins and multiprotein complexes, and
4. During exposure to stresses that cause previously folded proteins to unfold, becomes evident. In the few cases where folding has been studied both in vivo and in vitro, it appears that the folding pathways are similar.

Cells have solved the problem of misfolding and aggregation, to a considerable extent at least, through the participation of molecular chaperones in the in vivo folding process. Many investigations in the past few years have confirmed the critical role of molecular chaperones in protein folding in the cell. Although much has been learned about the function of chaperones in protein folding, and the general outline of the process is thought to be understood, there are still many important unresolved issues, and new chaperones and cochaperones are still being discovered.

The molecular chaperones involved in the folding of newly synthesized proteins recognize nonnative substrate proteins predominantly via their exposed hydrophobic residues. The major chaperone classes are 40-kDa heat shock protein (HSP40; the DnaJ family), 60-kDa heat shock protein [HSP60; including GroEL and the T-complex polypeptide 1 (TCP-1) ring complexes], 70-kDa heat shock protein (HSP70), and 90-kDa heat shock protein (HSP90). All these chaperones can prevent the aggregation of at least some unfolded proteins. For HSP60 and HSP70, their activity is modulated by the binding and hydrolysis of ATP. The HSP70 (DnaK in *Escherichia coli*) bind to nascent polypeptide chains on ribosomes, preventing their premature folding, misfolding, or aggregation, as well as to newly synthesized proteins in the process of translocation from the cytosol into the mitochondria and the endoplasmic reticulum (ER). The HSP70 are regulated by HSP40 (DnaJ or its homologs). The HSP60 are large oligomeric ring-shaped proteins known as chaperonins that bind partially folded intermediates, preventing their aggregation, and facilitating their folding and assembly. This family is composed of GroEL-like proteins in eubacteria, mitochondria, and chloroplasts and the TCP-1 (CCT or TRiC) family in the eukaryotic cytosol and the archaea. The HSP60 (GroEL in *E. coli*) are large, usually tetradecameric proteins with a central cavity in which nonnative protein structures bind.

The HSP60 are found in all biological compartments except the ER. The HSP60 are regulated by a cochaperone, chaperonin 10 (cpn10) (GroES in *E. coli*). In addition to preventing aggregation, it has been suggested that HSP60 may permit misfolded structures to unfold and refold. The HSP90 are associated with a number of proteins and play important roles in modulating their activity, most notably the steroid receptors. A number of other proteins involved in the folding of many newly synthesized proteins are often considered to be molecular chaperones; these include protein disulfide isomerase and peptidyl prolyl isomerase, which catalyze the rearrangement of disulfide bonds and isomerization of peptide bonds around Pro residues, respectively, and are perhaps better considered to be folding catalysts rather than chaperones. As mentioned previously, there are also a number of more specific chaperones that are involved in the folding/ assembly of only one, or a very limited number, of particular substrate proteins.

Chaperones are catalysts in the sense that they transiently interact with their substrate proteins but are not present in the final folded product, and also in that they increase the yield of folded protein. However, there is no good evidence that they actually enhance the spontaneous rate of folding itself, although they may appear to do this by minimizing off-pathway reactions.

A brief perusal of the literature demonstrates that our knowledge of molecular chaperones is growing at an enormous pace. To put these new discoveries in context, a few more general points are worthy of note. Although in many respects the field of molecular chaperones can now be considered a mature one, in that it has passed its first decade of life, and the broad outlines, at least, are reasonably well established, there are still many outstanding questions. Furthermore, there are many areas of considerable controversy, and many of these relate to fundamental questions.

For example, we do not yet know with certainty whether all newly synthesized proteins interact with chaperones, although it is likely that they do. We certainly do not know much about all the interactions between the various chaperones themselves, as well as with newly synthesized proteins or other chaperone target proteins. As discussed in this review, there are significant controversies concerning which chaperones interact first with nascent polypeptides, and even whether all nascent polypeptides interact with chaperones. The GroEL family of chaperones has been intensively studied, especially in the context of in vitro protein folding, yet it is not clear just how important a role this family (the cpn60 chaperonins and their TCP-1 eukaryotic homologs) play in the folding of most proteins in the cell.

We are only now beginning to get a picture of the apparently ubiquitous role of the HSP90 family in many critical processes in the cell, especially those involving protein-protein interactions. Recently, several new "accessory" proteins have been discovered, which apparently act as "cochaperones." Again, their significance to protein folding and denaturation in the cell in general is unclear at the present time; they may be highly specialized or may turn out to be critical in a broad range of cellular processes involving chaperones. Although some of the chaperones clearly are important in preventing protein aggregation, there is as yet no good evidence that chaperones play a role significant in the opposite side of this equation, namely, in solubilizing protein aggregates, although it would seem likely that this may in fact be a function of some chaperones. Even at the level of the specific mechanisms of chaperone function, there are many controversial aspects, and those in the field know there have been some quite rancorous discussions over competing mechanisms. Thus the molecular chaperone field is one in which there are still many outstanding questions, including some quite fundamental ones. Consequently, chaperone scientists are likely to remain busy for a long time to come.

Hsp90: A Specialized but Essential Protein-folding Tool

Cells are equipped with several classes of structurally unrelated molecular chaperones to ensure both the proper folding of proteins upon synthesis and their refolding under conditions of denaturing stress. Different chaperones follow distinct strategies to achieve the general goal of preventing protein misfolding and aggregation in the highly crowded cellular environment. For example, the monomeric Hsp70 chaperone proteins promiscuously recognize short hydrophobic peptide segments such as those exposed in nascent polypeptide chains.

In contrast, the so-called chaperonins form oligomeric double ring structures capable of enclosing entire proteins or protein domains to promote folding in a sequestered environment. Chaperone proteins often cooperate in folding: a polypeptide that is bound by Hsp70 at an early stage of folding can be transferred to a chaperonin to reach its final native state. For certain client polypeptides in the eukaryotic cytosol, an alternate pathway has been proposed in which binding to Hsp70 is followed by an interaction with the dimeric chaperone protein Hsp90. Recent advances in understanding the structure and function of Hsp90 now indicate that it uses a protein-folding strategy different from that of either Hsp70 or the chaperonins. Here, we discuss the distinct biochemical mechanisms by which Hsp90 handles its substrate polypeptides and the unique biological functions of this curious protein-folding tool.

Unlike Hsp70, eukaryotic cytosolic Hsp90 does not act generally in nascent protein folding. Hsp90 is distinguished from other chaperones in that most of its known substrates are signal transduction proteins, the classical examples being steroid hormone receptors and signaling kinases. Because Hsp90 is essential for maintaining the activity of numerous signaling proteins, it plays a key role in cellular signal transduction networks. At a molecular level, Hsp90 binds to substrate proteins, which are in a near native state and thus at a late stage of folding poised for activation by ligand binding or interaction with other factors.

In fulfilling its role, Hsp90 operates as part of a multichaperone machinery in the cytosol, which includes Hsp70 as mentioned above but also peptidyl-prolyl isomerases and other cochaperones. This complexity of function has made Hsp90 more difficult to investigate than the Hsp70 or chaperonin proteins. For example, Hsp90 has only been established recently as an ATP-dependent chaperone and the exact functions of the Hsp90 ATPase activity are still being worked out. In the following sections, we will first describe the structure and mechanism of the Hsp90 protein, then focus on its regulation by cochaperones, and finally discuss the biological contribution of Hsp90 to cellular processes including signal transduction, protein degradation, and morphological evolution.

Molecular Chaperones: Biology and Prospects for Pharmacological Intervention

Successful expression of polypeptide gene products requires transcription machinery to generate messenger ribonucleic acids and translation machinery to synthesize polypeptide chains but the protein product is functional, and thus the gene is effectively expressed, only when the polypeptide chain is folded into its native three-dimensional conformation, appropriately localized within or secreted from the cell, and, in many instances, properly assembled into multicomponent complexes. Although the amino acid sequence alone is sufficient to dictate the native conformation of small proteins in vitro most polypeptides would fail to fold efficiently in the highly concentrated, complex, cellular environment without the assistance of yet another type of machinery. This latter type of machinery involves the so-called molecular chaperones, i.e., proteins adapted to facilitate protein folding. The chaperone concept was first proposed by Ellis and a large, highly active field of research has since developed to investigate the functions and physiological implications of molecular chaperones. Major aspects of this field have been recently reviewed and many details have been compiled in a more comprehensive manner.

There is an inherent problem in the folding of nascent polypeptide chains. Whether synthesized on free ribosomes in the cytoplasm or on ribosomes associated with endoplasmic reticulum (ER) nascent protein chains emerge in a linear manner. Consequently, hydrophobic stretches and other potentially interactive sites must sometimes wait for downstream sequences to emerge before appropriate folding interactions can be established. Inappropriate interactions can readily lead to poorly reversible conformations and aggregations that reduce the efficiency of native folding reactions. Molecular chaperones reversibly interact with nascent chains to minimize offpathway interactions and increase the yield of native folded protein.

In addition to the folding of nascent polypeptide chains, other cellular processes involve protein folding that requires chaperone participation. One of these is the transport of previously synthesized proteins across cell membranes, as best understood for mitochondrial protein import. Many mitochondrial proteins are encoded by nuclear genes, synthesized on free ribosomes in the cytoplasm, and subsequently transported across one or both mitochondrial membranes. Proteins are translocated across the mitochondrial membranes in a linear manner as unfolded chains, and there are both cytoplasmic and mitochondrial chaperones that participate in the unfolding and refolding pathways.

Heat Shock Proteins and Cardiovascular Pathophysiology

Heat Shock Proteins and Cardiovascular Pathophysiology.In the eukaryotic cell an intrinsic mechanism is present providing the ability to defend itself against external stressors from various sources. This defence mechanism probably evolved from the presence of a group of chaperones, playing a crucial role in governing proper protein assembly, folding, and transport. Upregulation of the synthesis of a number of these proteins upon environmental stress establishes a unique defence system to maintain cellular protein homeostasis and to ensure survival of the cell. In the cardiovascular system this enhanced protein synthesis leads to a transient but powerful increase in tolerance to such endangering situations as ischemia, hypoxia, oxidative injury, and endotoxemia. These socalled heat shock proteins interfere with several physiological processes within several cell organelles and, for proper functioning, are translocated to different compartments following stress-induced synthesis. In this review we describe the physiological role of heat shock proteins and discuss their protective potential against various stress agents in the cardiovascular system.

Since the first report on the heat-induced appearance of chromosomal puffings in salivary gland tissue of *Drosophila busckii* in 1962, a new research domain has been intensively explored. This research resulted in the discovery of a large number of related proteins and their physiological role in many prokaryotic and eukaryotic organisms, tissues, and individual cells and at the level of subcellular structures. These proteins were originally called "heat shock proteins" because they were discovered in salivary glands and other tissues of *Drosophila melanogaster* recovering from a so-called transient sublethal heat shock, during which body temperature was increased ~5°C above normal body core temperature. Such a mild heat shock elicited a heat shock response, characterized by the synthesis of new heat shock proteins normally almost absent in tissues of adult animals and by an increased synthesis of constitutively present or cognate heat shock proteins. This event was followed by a transient increased tolerance to high, normally lethal temperatures (thermotolerance). Later it was found that not only the tolerance to enhanced temperature increases, but also the resistance toward other events like hypoxia, ischemia, inflammation, and exposure to such cellular toxins as heavy metals, endotoxins, and reactive oxygen species (cross-tolerance), all imposing serious stress upon tissues and their composing cells.

The increased resistance toward stressful events has been described in a great variety of organisms, organs, and tissues, including the heart. During the last years, more and more information has become available on the specific role of individual heat shock proteins and on the mechanisms of transient activation of gene transcription, leading to the enhanced cardiac synthesis of these proteins. A better insight has also been acquired in such subcellular targets as molecules or organelles that can be protected by heat shock proteins. Now the phase is reached that upregulation of heat shock protein synthesis is considered as a powerful physiological, endogenous route for protecting crucial cellular homeostatic mechanisms against disturbing external factors. Therefore, it is propagated that they provide a new therapeutic tool to protect the human heart during and after transient ischemic attacks.

Over the last 5 years comprehensive reviews have been published on the chaperone activity of heat shock proteins and on their role in specific tissues and pathological processes. The role of such individual heat shock proteins as the 70-kDa heat shock protein, small heat shock proteins and the heat shock factors has also been highlighted.

This survey concentrates on the transient protection of the cardiovascular system in various mammalian species when heat shock protein synthesis is upregulated. The present state of the art regarding the stimuli, leading to the heat shock response, the transduction pathways involved in the activation transcription of the various heat shock genes, and the cellular structures protected by heat shock proteins is presented. Such new experimental tools as transgenic animals and transfection techniques to enhance or inhibit the transcription of heat shock genes are discussed. Although the physiological role of heat shock proteins and their protective potential under pathophysiological circumstances have been studied in detail in animal models, knowledge of the role and effects of these proteins in humans is still limited. Where possible the relevance of the observations for the human situation is discussed.

Metallochaperones, an Intracellular Shuttle Service for Metal Ions

Enzymes that employ transition metals as co-factors are housed in a wide variety of intracellular locations or are exported to the extracellular milieu. A key question then arises: how are specific metal co-factors transported to diverse locations and subsequently sorted into the correct metalloenzymes? The mechanisms by which these tasks are accomplished are just now being unraveled. A new family of soluble metal receptor proteins known as "metallochaperones" is emerging that act in the intracellular trafficking of metal ions. Although transition elements can be quite toxic, these metal receptors are not detoxification proteins; they clearly function in a "chaperone-like" manner, guiding and protecting the metal ion while facilitating appropriate partnerships. Here we will review the most recent advances in our understanding of copper metallochaperones and discuss mechanisms that may be relevant to other essential, yet potentially toxic, metal ions.

Conjugated Proteins

Covalently bonded to prosthetic groups such as glycoprotein and metalloprotein.

Cytokines

Regulate immunity, inflammation, apoptosis, and hematopoiesis.

Interleukins

Interleukins are the cytokines that act specifically as mediators between leucocytes. The following table shows the major source and effects of various types of interleukins.

	Major source	Major effects
IL-1	Macrophages	Stimulation of T cells and antigen-presenting cells.B-cell growth and anibody production.Promotes hematopoiesis (blood cell formation).
IL-2	Activated T cells	Proliferation of activated T cells.
IL-3	T lymphocytes	Growth of blood cell precursors.
IL-4	T cells and mast cells	B-cell proliferation.IgE production.
IL-5	T cells and mast cells	Eosinophil growth.
IL-6	Activated T cells	Synergistic effects with IL-1 or TNFa.
IL-7	thymus and bone marrow stromal cells	Development of T cell and B cell precursors.
IL-8	Macrophages	Chemoattracts neutrophils.
IL-9	Activated T cells	Promotes growth of T cells and

		mast cells.
IL-10	Activated T cells, B cells and monocytes	Inhibits inflammatory and immune responses.
IL-11	Stromal cells	Synergistic effects on hematopoiesis.
IL-12	Macrophages, B cells	Promotes T_H1 cells while suppressing T_H2 functions
IL-13	T_H2 cells	Similar to IL-4 effects
IL-15	Epithelial cells and monocytes	Similar to IL-2 effects
IL-16	CD8 T cells	Chemoattracts CD4 T cells.
IL-17	Activated memory T cells	Promotes T cell proliferation.
IL-18	Macrophages	Induces IFNg production.

Most cytokines use the Jak-STAT pathway to transmit signals, but IL-1 uses a much more complex pathway (PNAS, 2001).

Interferons

Interferons are the cytokines that can "interfere" with viral growth. They also have the ability to inhibit proliferation and modulate immune responses. Four types of interferons have been identified: IFN-α, IFN-β, IFN-ω and IFN-γ. The first three are Type I IFNs which have relatively high antiviral potency. IFN-γ is the Type II IFN, also called immune IFN.

Type I IFNs are produced by macrophages, neutrophils and other somatic cells in response to infection by viruses or bacteria. After they are released, they may bind to their receptors that are expressed on most cell types, resulting in the production of over 30 different proteins in the target cell. Among them, two enzymes play a critical role in the inhibition of viral replication: RNA-dependent protein kinase (PKR) and 2'-5' oligoadenylate synthetase (2-5A synthetase). Both enzymes can be activated by double-stranded RNA which may be present in some viruses. The activated PKR can phosphorylate a protein (eIF2) to inhibit protein synthesis. The 2-5A synthetase produces oligoadenylate that can bind and activate a cellular endonuclease to degrade mRNA.

IFN-γ is produced in activated T_H1 and NK cells, particularly in response to IL-2 and IL-12. Its production is suppressed by IL-4, IL-10, and TGF-β. Binding of IFN-γ to its receptor increases the expression of class I MHC on all somatic cells. It also enhances the expression of class II MHC on antigen-presenting cells. IFN-γ may also activate macrophages, neutrophils and NK cells.

Tumor Necrosis Factors

Tumor necrosis factors are the cytokines produced mainly by macrophages and T lymphocytes that help regulate the immune response and hematopoiesis (blood cell formation). There are two types of TNF:

TNFα — also called cachectin, produced by macrophages.

TNFβ — also called lymphotoxin, produced by activated CD4+ T cells.

Human TNFα contains 157 amino acids, generated from a precursor protein of 233 amino acids by a converting enzyme called TACE. TNFβ contains 171 amino acids. Both TNFa and TNFb bind to the Type I and Type II receptors of the TNF receptor superfamily. The Type I receptor (TNFRI) has a molecular weight of 55kD while the Type II receptor (TNFRII) has 75kD. TNFRI contains a death domain which is involved in apoptosis. Most cytokines use the Jak-STAT pathway to transmit signals, but TNF may bind TNFRI to activate the NF-kB pathway.

TNF and IL-1 are structurally unrelated and bind different receptors, but their biological effects are quite similar. Both can stimulate lymphocyte proliferation and induce acute-phase responses.

Chemokines

Chemokines are the cytokines that may activate or chemoattract leukocytes. Each chemokine contains 65 ~ 120 amino acids, with molecular weight of 8 ~ 10 kD. Their receptors belong to G-protein-coupled receptors. Since the entry of HIV into host cells requires chemokine receptors, their antagonists are being developed to treat AIDS.

Hormones

Examples: insulin, growth hormone, and prolactin.

Prions

Prions are pathogenic variants of proteins that are naturally produced in nerve cells and certain other cells. The normal "healthy" prions are referred to as PrPc (Prion Protein cellular). The word "prion" stands for "proteinaceous infectious particle" and so should properly only be applied to the pathogenic variants. In this picture, the production of PrPc is illustrated from the nucleus at bottom right. RNA that codes for PrPc is produced in the nucleus and exits via the nuclear pore. The RNA then passes along ribosomes attached to the rER. PrPc is formed in the rER and then progresses up through the Golgi. At the upper face of the Golgi, vesicles containing PrPc bud off and travel to the cell surface. Here, they fuse with the cell membrane and so discharge their cargo at "PrPc". By this means, the cellular proteins come to sit on the exterior of the cell.

PrPc encounter rogue (purple) prions. These are termed PrPsc (sc stands for scrapie, the prion disease of sheep). The rogue prions seem to force the normal proteins to change shape. Both types of protein, the PrPc and their corresponding prions, are the same chemical - just different shapes. It is this shape-flipping which is equivalent to the transmission of infection. Such a conformational shift or flip could happen at the cell surface or in caveolae (one is shown as a small invagination in the cell membrane). Residual PrPc might continue to be flipped by contact with the rogue confomations for some time in these vesicles. Prions polymerise, finally appearing as purple fibrils in the picture at "P". PrPsc is resistant to degradation by the enzymes contained in the lysosomes that are seen floating nearby. Consequently, PrPsc accumulates in the cell. PrPsc vesicles may also travel to the Golgi and intercept PrPc that is being processed there. In this way, PrPc particles could be switched to the rogue form before they reach the surface of the cell. By such mechanisms, PrPc might be switched to PrPsc at various points in and on the cell.

Prions could enter the brain along the axons of neurons. This probably happens by a retrograde flow of prion filled vesicles. These are shown in the picture as purple spheres ascending the axon like elevators going up a shaft. Another route of entry could be the blood, probably in immune cells. A lymphocyte is shown exiting the capillary at bottom left where it could then contact the astrocyte. Astrocytes and other glial cells also support the production of prions.

Prions are implicated in BSE (Bovine Spongiform Encephalopathy or Mad Cow Disease) and its human counterpart nvCJD (new variant Creutzfeldt Jakob Disease). These and similar diseases are known as TSEs (Transmissible Spongiform Encephalopathies). TSEs afflict other species, most famously sheep (scrapie).

Structural Proteins

Collagen is the major insoluble fibrous protein in the extracellular matrix and in connective tissue. In fact, it is the single most abundant protein in the animal kingdom. There are at least 16 types of collagen, but 80 – 90 percent of the collagen in the body consists of types I, II, and III. These collagen molecules pack together to form long thin *fibrils* of similar structure. Type IV, in contrast, forms a two-dimensional reticulum; several other types associate with fibril-type collagens, linking them to each other or to other matrix components. At one time it was thought that all collagens were secreted by fibroblasts in connective tissue, but we now know that numerous epithelial cells make certain types of collagens. The various collagens and the structures they form all serve the same purpose, to help tissues withstand stretching.

Myosins are a large family of motor proteins found in eukaryotic tissues. They are responsible for actin-based motility.

Most myosin molecules are composed of both a head and a tail domain.

- The *head domain* binds the filamentous actin, and uses ATP hydrolysis to generate force and to "walk" along the filament towards the (+) end (with the exception of one family member, myosin VI, which moves towards the (-) end).

- The *tail domain* generally mediates interaction with cargo molecules and/or other myosin subunits.

Myosin I

Myosin I's function is unknown, but it is believed to be responsible for vessicle transport or the contraction vacuole of cells.[1]

Myosin II

Myosin II, responsible for skeletal muscle contraction, is perhaps the best-studied example of these properties.

- Myosin II contains two *heavy chains,* each about 2000 amino acids in length, which constitute the head and tail domains. Each of these heavy chains contains the N-terminal head domain, while the C-terminal tails take on a coiled-coil morphology, holding the two heavy chains together (imagine two snakes wrapped around each other, such as in a caduceus). Thus, myosin II has two heads.
- It also contains 4 *light chains* (2 per head), which bind the heavy chains in the "neck" region between the head and tail.

In muscle cells, it is myosin II that is responsible for producing the contractile force. Here, the long coiled-coil tails of the individual myosin molecules join together, forming the thick filaments of the sarcomere. The force-producing head domains stick out from the side of the thick filament, ready to walk along the adjacent actin-based thin filaments in response to the proper chemical signals.

Ubiquitin

The marker for protein degradation.If a protein binds to ubiquitin, it will be degraded by proteasome.

Like all macromolecular components of an organism, the proteome is in a dynamic state of synthesis and degradation.

During proteolysis, the peptide bonds that link amino acids are hydrolyzed, and free amino acids are released. The process is carried out by a diverse group of enzymes termed proteases. During proteolysis, the energy invested in the synthesis of the peptide bond is released. Distinct proteolytic mechanisms serve different physiological requirements and allow the organism to accommodate to changing environmental and pathophysiological conditions.

One should distinguish between destruction of "foreign" and "self" proteins. Foreign dietary proteins are degraded "outside" the body, in the lumen of the gastrointestinal tract. To avoid triggering an immune response, the epithelial lining of the digestive tract does not allow absorption of intact proteins into the body, and they are degraded to nonantigenic amino acids that are absorbed by the body and serve as building blocks for synthesis of its own proteins. Self proteins can also be classified into extracellular and intracellular; the two groups of proteins are degraded via two distinct mechanisms. Extracellular proteins such as the blood coagulation factors, immunoglobulins, albumin, cargo-carrying proteins [such as the core protein of the low-density lipoprotein (LDL)], and peptide hormones (such as insulin) are taken up via pinocytosis or receptor-mediated endocytosis. They are then carried via a series of vesicles (endosomes) that fuse with primary lysosomes where they are degraded. During this process, the extracellular proteins are never exposed to the intracellular environment (the cytosol) and remain "extracellular" (topologically) throughout. Degradation of proteins in lysosomes is not specific, and all engulfed proteins exposed to lysosomal proteases are degraded at approximately the same rate.

Several observations lead to the prediction that degradation of intracellular proteins must be carried out by completely distinct mechanisms. The process is highly specific, and different proteins have half-life times that vary from a few minutes to several days (e.g., the muscle proteins actin and myosin) and up to a few years (crystalline). Furthermore, inhibitors of lysosomal degradation, weak bases such as chloroquine, for example, do not have any effect on degradation of intracellular proteins under basal metabolic conditions. These compounds titrate the normal acidic intralysosomal pH and bring it to a point that does not allow activity of the lysosomal proteases. These findings led to the hypothesis that degradation of intracellular proteins must be carried out by a nonlysosomal proteolytic system that is endowed with a high degree of specificity toward its substrates. Also, the fact that the proteolytic enzymes and their substrates reside in the same cellular compartment predicted a requirement for tightly regulated machinery that uses metabolic energy for control. The discovery of the ubiquitin-proteasome proteolytic pathway has resolved these enigmas.

Degradation of a protein via the ubiquitin-proteasome pathway involves two discrete and successive steps:

1. Tagging of the substrate by covalent attachment of multiple ubiquitin molecules and
2. Degradation of the tagged protein by the 26S proteasome complex with release of free and reusable ubiquitin. This last process is mediated by ubiquitin recycling enzymes [deubiquitinating enzymes (DUBs)].

Conjugation of ubiquitin, a highly evolutionarily conserved 76-residue polypeptide, to the protein substrate proceeds via a three-step cascade mechanism. Initially, the ubiquitin-activating enzyme E1 activates ubiquitin in an ATP-requiring reaction to generate a high-energy thiol ester intermediate, E1-S~ubiquitin. One of several E2 enzymes [ubiquitin-carrier proteins or ubiquitin-conjugating enzymes (UBCs)] transfers the activated ubiquitin moiety from E1, via an additional high-energy thiol ester intermediate, E2-S~ubiquitin, to the substrate that is specifically bound to a member of the ubiquitin-protein ligase family, E3. There are a number of different classes of E3 enzymes. For the HECT (homologous to the E6-AP COOH terminus) domain E3s, the ubiquitin is transferred once again from the E2 enzyme to an active site Cys residue on the E3, to generate a third high-energy thiol ester intermediate, ubiquitin~ S-E3, before its transfer to the ligase-bound substrate. RING finger-containing E3s catalyze direct transfer of the activated ubiquitin moiety to the E3-bound substrate. E3s catalyze the last step in the conjugation process: covalent attachment of ubiquitin to the substrate. The ubiquitin molecule is generally transferred to an -NH_2 group of an internal Lys residue in the substrate to generate a covalent isopeptide bond. In some cases, however, ubiquitin is conjugated to the NH_2-terminal amino group of the substrate. By successively adding activated ubiquitin moieties to internal Lys residues on the previously conjugated ubiquitin molecule, a polyubiquitin chain is synthesized. The chain is recognized by the downstream 26S proteasome complex. Thus E3s play a key role in the ubiquitin-mediated proteolytic cascade since they serve as the specific recognition factors of the system.

The proteasome is a large, 26S, multicatalytic protease that degrades polyubiquitinated proteins to small peptides. It is composed of two subcomplexes: a 20S core particle (CP) that carries the catalytic activity and a regulatory 19S regulatory particle (RP). The 20S CP is a barrel-shaped structure composed of four stacked rings, two identical outerrings and two identical innerrings. The eukaryotic and rings are composed each of seven distinct subunits, giving the 20S complex the general structure of $_{1\text{-}71\text{-}71\text{-}71\text{-}7}$. The catalytic sites are localized to some of the subunits. Each extremity of the 20S barrel can be capped by a 19S RP. One important function of the 19S RP is to recognize ubiquitinated proteins and other potential substrates of the proteasome.A ubiquitin-binding subunit of the 19S RP has indeed been identified; however, its importance and mode of action have not been discerned. A second function of the 19S RP is to open an orifice in the ring that will allow entry of the substrate into the proteolytic chamber. Also, because a folded protein would not be able to fit through the narrow proteasomal channel, it is assumed that the 19S particle unfolds substrates and inserts them into the 20S CP. Both the channel opening function and the unfolding of the substrate require metabolic energy, and indeed, the 19S RP contains six different ATPase subunits. After degradation of the substrate, short peptides derived from the substrate are released, as well as reusable ubiquitin.

A major unresolved question is, How does the system achieve its high specificity and selectivity? Why are certain proteins extremely stable in the cell, whereas others are extremely short-lived? Why are certain proteins degraded only at a particular time point during the cell cycle or only after specific extracellular stimuli, yet they are stable under most other conditions? It appears that specificity of the ubiquitin system is determined by two distinct and unrelated groups of proteins:

E3s and ancillary proteins. First, within the ubiquitin system, substrates must be specifically recognized by an appropriate E3 as a prerequisite to their ubiquitination. In most cases however, substrates are not recognized in a constitutive manner and are not recognized directly by the E3. In some instances, the E3 must "be switched on" by undergoing posttranslational modification to yield an active form that recognizes the substrate. In many other cases, it is the substrate that must undergo a certain change that renders it susceptible for recognition.

The stability of additional proteins depends on association with ancillary proteins such as molecular chaperones that act as recognition elements in *trans* and serve as a link to the appropriate ligase. Others, such as certain transcription factors, have to dissociate from the specific DNA sequence to which they bind in order to be recognized by the system. Stability of yet other proteins depends on oligomerization. Thus, in addition to the E3s themselves, modifying enzymes (such as kinases), ancillary proteins, or DNA sequences to which substrates bind, also play an important role in the recognition process.

Ubiquitin-mediated proteolysis of a variety of cellular proteins plays an important role in many basic cellular processes. Among these are regulation of cell cycle and division, differentiation and development, involvement in the cellular response to stress and extracellular effectors, morphogenesis of neuronal networks, modulation of cell surface receptors, ion channels and the secretory pathway, DNA repair, transcriptional regulation, transcriptional silencing, long-term memory, circadian rhythms, regulation of the immune and inflammatory responses, and biogenesis of organelles. The list of cellular proteins that are targeted by ubiquitin is growing rapidly. Among them are cell cycle regulators such as cyclins, cyclin-dependent kinase inhibitors, and proteins involved in sister chromatid separation, tumor suppressors, as well as transcriptional activators and their inhibitors. Cell surface receptors and endoplasmic reticulum (ER) proteins are also targeted by the system. Finally, mutated and denatured/misfolded proteins are recognized specifically and are removed efficiently. In this capacity, the system is a key player in the cellular quality control and defence mechanisms.

With the numerous substrate proteins targeted and the multitude of processes involved, it is not surprising that aberrations in the ubiquitin system have been implicated in the pathogenesis of many inherited and acquired human pathologies. In some cases, the linkage between the proteolytic system and the resulting pathology is direct, whereas in others it is less obvious. It is impossible to cover in a single review, or even in a comprehensive monograph, all that we currently know of the ubiquitin system, its enzymatic components, ancillary proteins, modes of action, mechanisms of substrate recognition, modes of regulation, and most of all, the numerous processes it is involved in or the consequences of aberrations in its activity.

We therefore decided to bring to the reader an updated view of the general components of the system, the conjugating enzymes and the proteasome, explain their mode of action, and highlight, via specific examples, several of the processes in which the system is involved. We will end by describing in greater detail the recently evolving field of the involvement of this system in pathogenesis of human diseases. For more details about the ubiquitin system, the reader is referred to the many specific reviews written on it recently.

Chapter 3

Nucleic Acids, Genomes and Proteomics

A nucleotide is composed of three parts: pentose, base and phosphate group. In DNA or RNA, a pentose is associated with only one phosphate group, but a cellular free nucleotide (such as ATP) may contain more than one phosphate group. If all phosphate groups are removed, a nucleotide becomes a nucleoside.

PENTOSE

There are five different bases, each is denoted by a single letter as given in the parenthesis:

Adenine (A), Cytosine (C), Guanine (G), Thymine (T), and Uracil (U).

Among them,

A, C, G and T exist in DNA;

A, C, G and U exist in RNA.

Their chemical structures are shown in the following figure. A and G contain a pair of fused rings, classified as purines. C, T, and U contain only one ring, classified as pyrimidines.

We note that the chemical structure of uracil is simpler than thymine. This gives a good reason why RNA uses uracil, instead of thymine. However, why does DNA use thymine? The major requirement in designing DNA bases is that they must be able to form pairs. But adenine can pair with uracil as perfectly as with thymine. Then, why DNA chooses a more complex base? This question had been puzzling researchers for many years, until they understood the DNA repairing mechanism.

CELLULAR NUCLEOTIDES AND NUCLEOSIDES

In cells, a free nucleotide may contain one, two or three phosphate groups. The energy carrier ATP (adenosine triphosphate) has three phosphate groups; ADP (adenosine diphosphate) has two; AMP (adenosine monophosphate) has one.

If all phosphate groups are removed, a nucleotide becomes a nucleoside such as adenosine. The following table lists various nucleosides and nucleotides.

Table. Cellular nucleosides and nucleotides

Base	**Nucleoside (=base + pentose)**		**Nucleotide (= nucleoside + phosphate)**		
	Ribo-nucleoside	Deoxyribo-nucleoside	NMP dNMP	NDP dNDP	NTP dNTP
Purines					
Adenine	Adenosine	Deoxyadenosine	AMP	ADP	ATP
		dAMP	dADP	dATP	
Guanine	Guanosine	Deoxyguanosine	GMP	GDP	GTP
			dGMP	dGDP	dGTP
Pyrimidines					
Cytosine	Cytidine	Deoxycytidine	CMP	CDP	CTP
			dCMP	dCDP	dCTP
Thymine	Thymidine	Deoxythymidine	TMP	TDP	TTP
		dTMP	dTDP	dTTP	
Uracil	Uridine	Deoxyuridine	UMP	UDP	UTP
			dUMP	dUDP	dUTP

NMP = nucleoside monophosphate

NDP = nucleoside diphosphate

NTP = nucleoside triphosphate

dNMP = deoxynucleoside monophosphate

dNDP = deoxynucleoside diphosphate

dNTP = deoxynucleoside triphosphate

(nucleoside = ribonucleoside; deoxynucleoside = deoxyribonucleoside)

THE NUCLEIC ACID CHAIN

In a nucleic acid chain, two nucleotides are linked by a phosphodiester bond, which may be formed by the condensation reaction similar to the formation of the peptide bond. In cells, such process has been found in the ligation between two nucleic acid fragments. However, the whole nucleic acid chain is usually synthesized by RNA polymerase or DNA polymerase.

Like peptide chains, a nucleic acid chain also has orientation: its 5' end contains a free phosphate group and 3' end contains a free hydroxyl group. Synthesis of a nucleic acid chain always proceeds from 5' to 3'. Therefore, unless specified otherwise, the sequence of a nucleic acid chain is written from 5' to 3' (left to right).

In DNA or RNA, a nucleic acid chain is also called a strand. A DNA molecule typically contains two strands whereas most RNA molecules contain a single strand.

The length of a nucleic acid chain is represented by the number of bases. In the case of a double-stranded nucleic acid, bases are paired between two strands. Therefore, its length is given by the number of base pairs (bp). 1 kb = 1000 bases or bp; 1 Mb = 1 million bases or bp. Oligonucleotides refer to short nucleic acid chains (< 50 bases or bp) and polynucleotides have longer chains.

STRUCTURE OF DNA

A DNA molecule has two strands, held together by the hydrogen bonding between their bases. As shown in the above figure, adenine can form two hydrogen bonds with thymine; cytosine can form three hydrogen bonds with guanine. Although other base pairs [e.g., (G:T) and (C:T)] may also form hydrogen bonds, their strengths are not as strong as (C:G) and (A:T) found in natural DNA molecules.

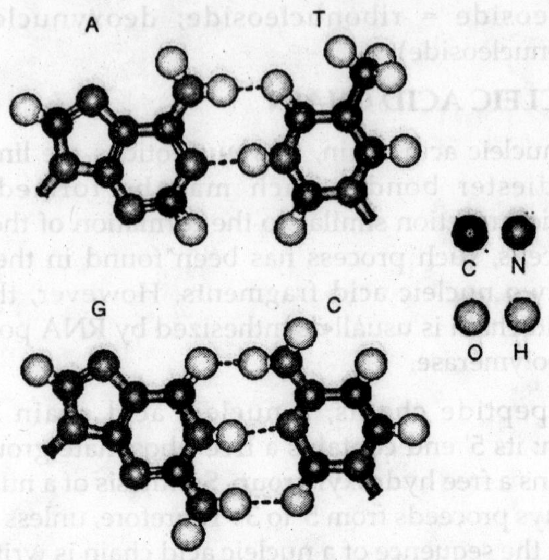

Fig. Computer model of base pairing in DNA. In a normal DNA molecule, adenine (A) is paired with thymine (T), guanine (G) is paired with cytosine (C). The uracil (U) of RNA can also pair with adenine (A), since U differs from T by only a methyl group located on the other side of hydrogen bonding.

Due to the specific base pairing, DNA's two strands are complementary to each other. Hence, the nucleotide sequence of one strand determines the sequence of another strand. For example the sequence of the two strands can be written as

5' -ACT- 3'

3' -TGA- 5'

Note that they obey the (A:T) and (C:G) pairing rule. If we know the sequence of one strand, we can deduce the sequence of another strand. For this reason, a DNA database needs to store only the sequence of one strand. By convention, the sequence in a DNA database refers to the sequence of the 5' to 3' strand (left to right).

DNA'S B FORM, A FORM AND Z FORM

In a DNA molecule, the two strands are not parallel, but intertwined with each other. Each strand looks like α helix. The two strands form a "double helix" structure, which was first discovered by James D. Watson and Francis Crick in 1953. In this structure, also known as the B form, the helix makes a turn every 3.4 nm, and the distance between two neighboring base pairs is 0.34 nm. Hence, there are about 10 pairs per turn. The intertwined strands make two grooves of different widths, referred to as the major groove and the minor groove, which may facilitate binding with specific proteins.

In a solution with higher salt concentrations or with alcohol added, the DNA structure may change to an A form, which is still right-handed, but every 2.3 nm makes a turn and there are 11 base pairs per turn.

Another DNA structure is called the Z form, because its bases seem to zigzag. Z DNA is left-handed. One turn spans 4.6 nm, comprising 12 base pairs. The DNA molecule with alternating G-C sequences in alcohol or high salt solution tends to have such structure.

RNA STRUCTURE AND FUNCTION

Most cellular RNA molecules are single stranded. They may form secondary structures such as stem-loop and hairpin.

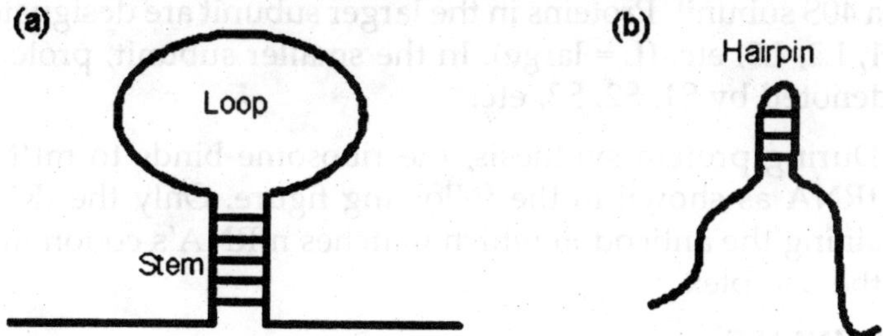

Fig. Secondary structure of RNA. (a) stem-loop. (b) hairpin.

The major role of RNA is to participate in protein synthesis, which requires three classes of RNA

MESSENGER RNA (MRNA)

mRNA is transcribed from DNA, carrying information for protein synthesis. Three consecutive nucleotides in mRNA encode an amino acid or a stop signal for protein synthesis. The trinucleotide is know as a codon.

TRANSFER RNA (TRNA)

The major role of tRNA is to translate mRNA sequence into amino acid sequence. A tRNA molecule consists of 70-80 nucleotides. Its secondary and tertiary structures are shown in figures. Some nucleotides in tRNA have been modified, such as dihydrouridine (D),pseudouridine (Y), and inosine (I). In dihydrouridine, a hydrogen atom is added to each C5 and C6 of uracil. In pseudouridine, the ribose is attached to C5, instead of the normal N1. Inosine plays an important role in codon recognition. In addition to these modifications, a few nucleosides are methylated.

RIBOSOMAL RNA (RRNA)

In prokaryotes, the ribosomal RNA (rRNA) has three types: 23S, 5S, and 16S. In mammals, four types of rRNA have been found: 28S, 5.8S, 5S and 18S. The unit "S" stands for Svedberg, which is a measure of the sedimentation rate. After rRNA molecules are produced in the nucleus, they are transported to the cytoplasm, where they combine with tens of specific proteins to form a ribosome. In prokaryotes, the size of a ribosome is 70S, consisting of two subunits: 50S and 30S. The size of a mammalian ribosome is 80S, comprising a 60S and a 40S subunit. Proteins in the larger subunit are designated as L1, L2, L3, etc. (L = large). In the smaller subunit, proteins are denoted by S1, S2, S3, etc.

During protein synthesis, the ribosome binds to mRNA and tRNA as shown in the following figure. Only the tRNA containing the anticodon which matches mRNA's codon may join the complex.

RIBOZYMES

The RNA molecules with catalytic activity. The VS ribozyme is one of the class of nucleolytic ribozymes that includes the hammerhead, hairpin, and hepatitis delta virus ribozymes. These carry out reversible cleavage and ligation reactions at a specific site by transesterification reactions involving 2'- and 5'-oxygen and the 3'-phosphorus atoms. The Varkud satellite (VS) RNA is an abundant transcript from DNA found in the mitochondria of a number of natural isolates of *Neurospora*. Collins and coworkers found that the VS RNA contains an element capable of self-cleavage which is thought to act in the processing of replication intermediates. At ~150 nucleotides, the VS ribozyme is the largest of the nucleolytic ribozymes, and is now the only member of this group for which there is no crystal structure.

In terms of sequence, the VS ribozyme has nothing in common with the other nucleolytic ribozymes. Virtually all of the early work on this ribozyme was carried out in the Collins laboratory. The secondary structure was deduced by a combination of probing and mutagenesis. A 5' stem–loop structure (helix I) contains an internal loop in which the ribozyme cleavage occurs. This is linked to an H-shaped structure comprising helices II through VI. Stem–loop I can be physically disconnected from the rest of the RNA, to create a ribozyme (helices II–VI) that acts in *trans* on the substrate. The VS ribozyme is unique among the nucleolytic ribozymes in that the interaction between the ribozyme and its substrate involves essentially tertiary contacts. In this aspect, it resembles more the ribozymes RNaseP which cleaves pre-tRNAs, and group I self-splicing introns, which bind and cleave a short splice-site duplex RNA. Part of this interaction is the formation of loop–loop pairing between the GAC triplet in the terminal loop of stem V, and a GUC triplet in the substrate. The latter is largely contained within the terminal loop, but exposure of the 3' cytosine would require the opening of the terminal base pair, and it is proposed that this results in a change in the conformation of the substrate that is important for the cleavage reaction. In its *trans*-acting form, the ribozyme is very well behaved, especially if base pairing in helix Ia is absent. The progress of the cleavage reaction is well fitted by a single exponential, and the kinetics conform to an enzyme-type scheme involving formation of a noncovalent complex followed by the chemical step, with values of $K_d = 1\ \mu M$ and $k_2 = 2\ min^{-1}$

The VS ribozyme is the largest nucleolytic ribozyme, for which there is no crystal structure to date. The ribozyme consists of five helical sections, organized by two three-way junctions. The global structure has been determined by solution methods, particularly FRET. The substrate stem–loop binds into a cleft formed between two helices, while making a loop–loop contact with another section of the ribozyme. The scissile phosphate makes a close contact with an internal loop (the A730 loop), the probable active site of the ribozyme. This loop contains a particularly critical nucleotide A756. Most changes to this nucleotide lead to three-orders of magnitude slower cleavage, and the Watson-Crick edge is especially important. NAIM experiments indicate that a protonated base is required at this position for the ligation reaction. A756 is thus a strong candidate for nucleobase participation in the catalytic chemistry.

SMALL RNA MOLECULES

RNA interference and other functions.

Major types of small RNA molecules:

- Small nuclear RNA (snRNA): involved in mRNA splicing.
- Small nucleolar RNA (snoRNA): Directs the modification of ribosomal RNAs.
- Micro-RNA (miRNA): Regulates gene expression. miRNA is only 20-25 nucleotides long. It may be divided into two subtypes: small interfering RNA (siRNA) and small temporally regulated RNA (stRNA). siRNA mediates RNA interference (RNAi) which has been widely used in functional genomics and also has therapeutic potential.

CHROMATIN STRUCTURE

Chromatin is the substance which becomes visible chromosomes during cell division. Its basic unit is nucleosome, composed of 146 bp DNA and eight histone proteins. The structure of chromatin is dynamically changing, at least in part, depending on the need of transcription. In the metaphase of cell division, the chromatin is condensed into the visible chromosome. At other times, the chromatin is less condensed, with some regions in a "Beads-On-a-String" conformation.

HISTONES AND NUCLEOSOMES

Histones are the proteins closely associated with DNA molecules. They are responsible for the structure of chromatin and play important roles in the regulation of gene expression. Five types of histones have been identified: H1 (or H5), H2A, H2B, H3 and H4. H1 and its homologous protein H5 are involved in higher-order structures of chromatin. The other four types of histones associate with DNA to form nucleosomes. H1 (or H5) has about 220 residues. Other types of histones are smaller, each consisting of 100-150 residues.

An important feature about histones is that they contain a few lysine (K) residues at the N terminus. Under normal cellular conditions, the R group of lysine is positively charged, which can interact with the negatively charged phosphates in DNA. The positive R group of lysine may be neutralized by acetylation, reducing the binding force between histones and DNA. Such mechanism has been demonstrated to play a major role in the regulation of gene transcription.

CONFORMATIONS OF CHROMATIN FIBERS

The structure of chromatin fibers is dynamically changing. In the condensed state, each nucleosome is associated with an H1 (or H5) to form a solenoid structure. H1 and H5 are called linker histones. They are essential in stabilizing the solenoid conformation. This can be demonstrated by extracting the condensed chromatin at a low salt concentration, which will release H1 or H5. Then, the "beads-on-a-string" conformation may be observed by the atomic force microscopy.

THE GENETIC CODE

Protein synthesis is based on the sequence of mRNA, which is made up of nucleotides while proteins are made up of amino acids. There must be a specific relationship between the nucleotide sequence and amino acid sequence. This relationship is the so called genetic code, which was deciphered by Marshall Nirenberg and his colleagues in early 1960s. One of their approaches is illustrated in figure. It turns out that three nucleotides (a codon) code for one amino acid.

Synthesis of a peptide always starts from methionine (Met), coded by AUG. The stop codon (UAA, UAG or UGA) signals the end of a peptide. This table applies to mRNA sequences. For DNA, U (uracil) should be replaced by T (thymine). In a DNA molecule, the sequence from an initiating codon (ATG) to a stop codon (TAA, TAG or TGA) is called an open reading frame (ORF), which is likely (but not always) to encode a protein or polypeptide.

CRACKING THE GENETIC CODE

1. Synthesize a trinucleotide (e.g. UUU) which mimics a codon in mRNA.
2. Prepare various types of aminoacyl-tRNA, e.g., Thr-tRNA, Phe-tRNA, Lys-tRNA, etc.
3. Radioactively label an aminoacyl-tRNA (e.g. Phe-tRNA) which might contain the anticodon for the synthesized trinucleotide.
4. Place the trinucleotide, aminoacyl-tRNA and ribosome on a nitrocellulose filter.

Individual trinucleotide and aminoacyl-tRNA can pass through the filter, but the ribosome is too big to pass through. Therefore, if the labeled aminoacyl-tRNA contains the anticodon for the trinucleotide, it will bind to the trinucleotide and ribosome on the filter. In this case, the radioactivity can be detected on the filter and the amino acid in the labeled aminoacyl-tRNA is likely to be encoded by the trinucleotide. If no radioactivity was detected, the trinucleotide is unlikely to be the codon of the amino acid. Most of the 64 possible codons can be determined by repeating this procedure for different trinucleotides and labellings.

MRNA AND DNA

mRNA is transcribed from DNA, carrying information for protein synthesis. Three consecutive nucleotides in mRNA encode an amino acid or a stop signal for protein synthesis. The trinucleotide is know as a codon.

ORDER IN THE GENETIC CODE

The genetic code is not randomly assigned. If an amino acid is coded by several codons, they often share the same sequence in the first two positions and differ in the third position. Such assignment is accomplished by the design of wobble position, but "*the evolutionary dynamic that shaped the code remains a mystery*".

From the above figure, we see that the amino acids with close physical properties have similar codons. For example, both Asp and Glu are negatively charged. Their codons all contain "GA" in the first two positions. The physical properties of Ser and Thr are also very close. They are coded by UCN and ACN (N = any), respectively.

One advantage for this assignment is to reduce the damage caused by replication errors or other mutations. For instance, if the third nucleotide of Ser codon is mutated, the produced amino acid remains Ser. If the first nucleotide of Ser codon is mutated to A, it will produce Thr, which is similar to Ser. Such mutation will not have significant effect on the protein's structure and function. Let us consider a hypothetical case that codons are randomly assigned. Suppose Ser were coded by GAG, GUA, GCU and UGG; Thr were coded by AAA, AGU, UUU and CCC. Then, any mutation would produce an amino acid different from Ser and Thr.

NON-STANDARD GENETIC CODES

The standard genetic code applies to most, but not all, cases. Exceptions have been found in the mitochondrial DNA of many organisms and in the nuclear DNA of a few lower organisms. Some examples are given in the following table.

Table. Examples of non-standard genetic codes.

Location	Species	Codon
Mitochondria	vertebrates	UGA→Trp; AGA,AGG→Stop
	Invertebrates	UGA→Trp; AGA,AGG→Ser
	Yeast	UGA→Trp; CUA, CUU, CUC, CUG→Thr
	Protista	UGA→Trp
Nucleus	Bacteria	GUG, UUG, AUU, CUG→Initiation
	Yeast	CUG→Ser
	Ciliates	UAA, UAG→Gln

Plant cells use the standard genetic code in both mitochondria and the nucleus.

GENES

By definition, a gene includes the entire nucleic acid sequence necessary for the expression of its product (peptide or RNA). Such sequence may be divided into regulatory region and transcriptional region. The regulatory region could be near or far from the transcriptional region. The transcriptional region consists of exons and introns. Exons encode a peptide or functional RNA. Introns will be removed after transcription.

As shown in the following figure, a typical DNA molecule consists of genes, pseudogenes and extragenic region. Pseudogenes are nonfunctional genes. They often originate from mutation of duplicated genes. Because duplicated genes have several copies, the organism can still survive even if a couple of them become nonfunctional.

Gene Family

"Gene family" refers to a set of genes with homologous sequences. For example, *H2A*, *H2B*, *H3* and *H4* are in the same histone gene family. Their products have similar structures and functions. Another example is the b-globin gene family located on the chromosome 11.

Duplicated Genes

Most proteins do not need duplicated genes, because the mRNA molecule transcribed from one gene can be used to produce many identical protein molecules. However, rRNA and tRNA are the final gene products. In order to accelerate the production process, all species contain an array of tandemly repeated RNA genes. The number of repeats ranges from tens to 24,000.

There are four types of rRNA in mammalian cells: 28S, 5.8S, 5S and 18S. In the human genome, 28S, 5.8S and 18S are clustered together. They form a single transcription unit which will be separated by specific enzymes after transcription. " Pre-rRNA" refers to their precursor. In humans, a repeat unit for the pre-rRNA has about 40 kb in length, including a 13-kb transcription unit and a 27-kb untranscribed spacer region. The transcription unit contains three spacers: ETS, ITS1 and ITS2. They will be removed during RNA processing.

REPETITIVE DNA SEQUENCES

A stretch of DNA sequence often repeats several times in the total DNA of a cell. For example, the following DNA sequence is just a small part of telomere located at the ends of each human chromosome:

```
...GGTTAGGGTTAGGGTTAGGGTTAGGGTTAGGGTTA
GGGTTAGGGTTAGGGTTAGGGTTAGGGTTAGGGTTAGG
GTTAGGGTTAGGGTTAGGGTTAGGGTTAGGGTTAG...
```

An entire telomere, about 15 kb, is constituted by thousands of the repeated sequence "GGGTTA".

DNA reassociation kinetics is used to classify the repetitive DNA sequences. The total DNA is first randomly cleaved into fragments with an average size of about 1000 bp. Then, they are heated to separate the complementary strands of each fragment. Subsequently, temperature is reduced to allow strand reassociation. If a fragment contains a sequence which is repeated many times in the total DNA, it will have greater chance to find a complementary strand and reassociate more quickly than other fragments with less repetitive sequences.

Based on the reassociation rate, DNA sequences are divided into three classes:

Highly repetitive: About 10-15per cent of mammalian DNA fragments reassociate very rapidly. This class includes tandem repeats.

Moderately repetitive: Roughly 25-40 per cent of mammalian DNA fragments reassociate at an intermediate rate. This class includes interspersed repeats (also known as mobile elements or transposable elements).

Single copy (or very low copy number): This class accounts for 50-60per cent of mammalian DNA

TANDEM REPEATS

Tandem repeats are an array of consecutive repeats. They include three subclasses: satellites, minisatellites and microsatellites. The name "satellites" comes from their optical spectra.

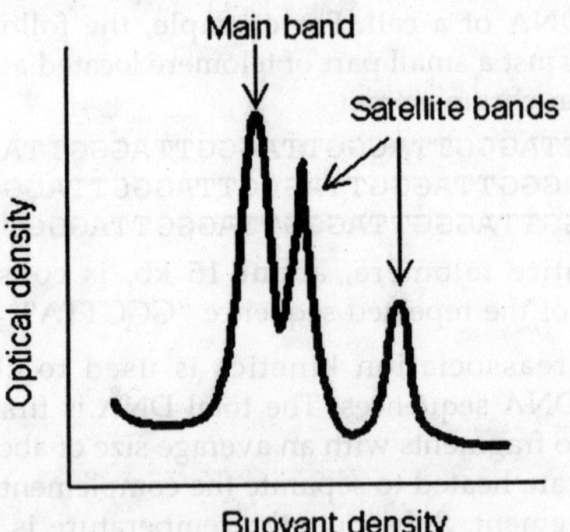

Fig. Illustration of Satellite Bands.

By using buoyant density gradient centrifugation, DNA fragments with significantly different base compositions may be separated, and then monitored by the absorption spectra of ultraviolet light. The main band represents the bulk DNA, and the "satellite" bands originate from tandem repeats.

Satellites

The size of a satellite DNA ranges from 100 kb to over 1 Mb. In humans, a well known example is the alphoid DNA located at the centromere of all chromosomes. Its repeat unit is 171 bp and the repetitive region accounts for 3-5per cent of the DNA in each chromosome. Other satellites have a shorter repeat unit. Most satellites in humans or in other organisms are located at the centromere.

Minisatellites

The size of a minisatellite ranges from 1 kb to 20 kb. One type of minisatellites is called variable number of tandem repeats (VNTR). Its repeat unit ranges from 9 bp to 80 bp. They are located in non-coding regions. The number of repeats for a given minisatellite may differ between individuals. This feature is the basis of DNA fingerprinting.

Another type of minisatellites is the telomere. In a human germ cell, the size of a telomere is about 15 kb. In an aging somatic cell, the telomere is shorter. The telomere contains tandemly repeated sequence GGGTTA.

Microsatellites

Microsatellites are also known as short tandem repeats (STR), because a repeat unit consists of only 1 to 6 bp and the whole repetitive region spans less than 150 bp. Similar to minisatellites, the number of repeats for a given microsatellite may differ between individuals. Therefore, microsatellites can also be used for DNA fingerprinting. In addition, both microsatellite and minisatellite patterns can provide information about paternity.

The term "polymorphism" describes the existence of different forms within a population, e.g., difference in the number of tandem repeats. All tandem repeat polymorphisms could result from DNA recombination during meiosis. The microsatellite polymorphism could also be caused by replication slippage.

INTERSPERSED REPEATS

Interspersed repeats are repeated DNA sequences located at dispersed regions in a genome. They are also known as mobile elements or transposable elements. As discussed a stretch of DNA sequence may be copied to a different location through DNA recombination. After many generations, such sequence (the repeat unit) could spread over various regions. Mobile elements were first discovered by Barbara McClintock in 1940s from the studies of corn. Subsequently, they were found in all kinds of organisms. In mammals, the most common mobile elements are LINEs and SINEs.

LINEs

LINEs stands for Long Interspersed Nuclear Elements. Its basic organization is shown below.

ORF1 and ORF2 are open reading frames. The protein product of ORF1 is called p40, with unknown function. ORF2 encodes a reverse transcriptase which is necessary for copying the element to other locations. The red colour regions on both ends are direct repeats. All mobile elements contain direct repeats whose origin is explained.

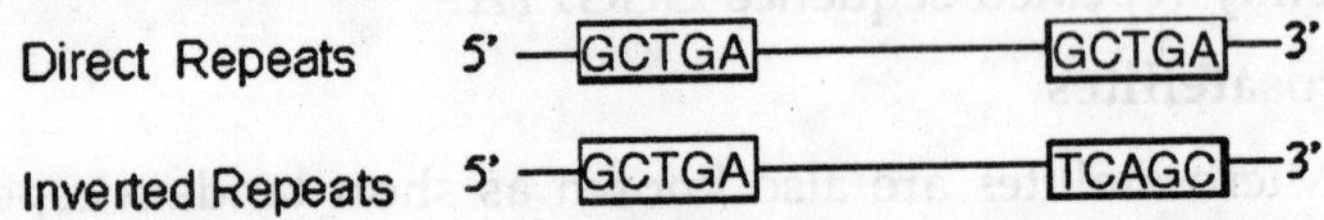

Fig. Comparison Between Direct Repeats and Inverted Repeats.

Direct repeats have the same sequence along a given direction. Inverted repeats have complementary sequences along opposite directions.

The most common LINEs in humans is the L1 family. A human genome contains about 60,000 to 100,1000 L1 elements.

SINEs

"SINEs" stands for Short Interspersed Nuclear Elements. Its length is about 300 bp. In humans, the most abundant SINEs is the *Alu* family. A human genome contains about 700,000 to 1,000,000 *Alu* sites. Although most LINEs and SINEs are located in extragenic regions, some of them are located in introns. For example, the human retinoblastoma gene (*RB* gene) is as long as 180 kb, consisting of 27 exons. Its introns contains many *Alu* and a few L1 elemtns.

JEFFERSON, ALLEGED SONS AND MICROSATELLITES

It has long been suggested that President Thomas Jefferson may be the father of Eston Hemings, the last child of Jefferson's slave Sally Hemings. Members of Woodson family also claimed that they were the descendants of President Jefferson. The following shows the number of repeats in each of 11 microsatellites obtained from the Y chromosomes of their male-line descendants. President Jefferson had no surviving sons. Therefore, his uncle's descendants were used in this study.

Descendants of President Jefferson's uncle:

15, 12, 4, 11, 3, 9, 11, 10, 15, 13, 7.

Descendants of Eston Hemings:

15, 12, 4, 11, 3, 9, 11, 10, 15, 13, 7.

Descendants of Thomas Woodson:

14, 12, 5, 11, 3, 10, 11, 13, 13, 13, 7.

This result indicates that Eston Hemings and Jefferson are genetically related, but Thomas Woodson's father was not a Jefferson.

GENOMES AND PROTEOMICS

GENOMES

"Genome"is the total genetic information of an organism. For most organisms, it is the complete DNA sequence. For RNA viruses, the genome is the complete RNA sequence, since their genetic information is encoded in RNA.

Table. The Genomes of Prominent Organisms.

Organism	Genome Size (Mb)	Gene Number
Hepatitis D virus	0.0017	2
Hepatitis B virus	0.0032	4
HIV-1	0.0092	9
Bacteriophage l	0.0485	90
Escherichia coli	4.6	4437
S. cerevisiae (yeast)	12	6300
C. elegans (nematode)	97	19000
D. melanogaster (fruit fly)	137	14000
Mus musculus (mouse)	3000	20000-30000
Homo sapiens (human)	3000	20000-30000

* 1 Mb = 1 million base pairs (for double-stranded DNA or RNA) or 1 million bases (for single-stranded DNA or RNA)

THE GENOMES OF MITOCHONDRIA AND CHLOROPLASTS

Mitochondria and chloroplasts (a type of plastid) are membrane-bound organelles that convert energy from foodstuffs (mitochondria) or sunlight (chloroplasts) into forms that can be used by the cell. The organelles likely evolved from bacteria that were endocytosed more than one billion years ago. Although the organelles maintain their own genomes, many genes encoding mitochondrial and chloroplast proteins are found in the cell nucleus.

Mitochondria are small, oval shaped organelles surrounded by two highly specialized membranes. Mitochondria are the sites of aerobic respiration, and are generally the major energy production centre in eukaryotes. Animal mitochondrial genomes are normally circular, ~16 kB in length, and encode 13 proteins used for energy production, as well as 22 tRNAs and 2 rRNAs. Plant mitochondrial genomes tend to be 10 - 150 times larger and contain additional genes. Many organisms use one genetic code to translate nuclear mRNAs, and a second one for their mitochondrial mRNAs.

Chloroplasts are larger than mitochondria, and are surrounded by two specialized membranes. In plants and some other eukaryotes, chloroplasts are the sites of photosynthesis, a process in which atmospheric carbon dioxide is "fixed" into organic compounds, and oxygen is released into the atmosphere. Chloroplast genomes are ~120 - 200 kB in length. Their 120 genes encode ribosomal RNAs and proteins, tRNAs, and proteins involved in photosynthesis. Chloroplast mRNAs are translated with the standard genetic code, although they often undergo extensive RNA editing, so it is difficult to predict the protein translations from genomic sequence.

Chapter 4

Gene Transcription

OVERVIEW OF GENE EXPRESSION

An organism may contain many types of somatic cells, each with distinct shape and function. However, they all have the same genome. The genes in a genome do not have any effect on cellular functions until they are "expressed". Different types of cells express different sets of genes, thereby exhibiting various shapes and functions.

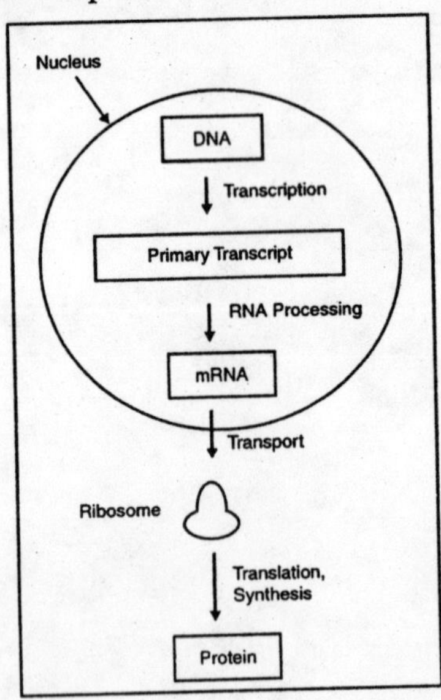

Fig. Essential Steps Involved in the Expression of Protein Genes.

"Gene expression" means the production of a protein or a functional RNA from its gene. Several steps are required:

Transcription: A DNA strand is used as a template to synthesize a complementary RNA strand, which is called the primary transcript.

RNA processing: This step involves modifications of the primary transcript to generate a mature mRNA (for protein genes) or a functional tRNA or rRNA.

For RNA genes (tRNA and rRNA), the expression is complete after a functional tRNA or rRNA is generated. However, protein genes require additional steps:

Nuclear transport: mRNA has to be transported from the nucleus to the cytoplasm for protein synthesis.

Protein synthesis: In the cytoplasm, mRNA binds to ribosomes, which can synthesize a polypeptide based on the sequence of mRNA.

THE CENTRAL DOGMA

According to the above process, the flow of genetic information is in the following direction:

DNA > RNA > Protein.

This rule was dubbed the "central dogma", because it was thought that the same principle would apply to all organisms. However, we now know that for RNA viruses, the flow of genetic information starts from RNA

OVERVIEW OF TRANSCRIPTION

Transcription is a process in which one DNA strand is used as template to synthesize a complementary RNA. The following is an example:

Note that uracil (U) of RNA is paired with adenine (A) of DNA. There are a few different names for these nucleic acid strands. The DNA strand which serves as the template may be called "template strand", "minus strand", or "antisense strand". The other DNA strand may be termed "non-template strand", "coding strand", "plus strand", or "sense strand". Since both DNA coding strand and RNA strand are complementary to the template strand, they have the same sequences except that T in the DNA coding strand is replaced by U in the RNA strand.

Growth of a nucleic acid strand is always in the 5' to 3' direction. This is true not only for the synthesis of RNA during transcription, but also for the synthesis of DNA during replication. The enzymes, called polymerases, are used to catalyze the synthesis of nucleic acid strands. RNA strands are synthesized by RNA polymerases. DNA strands are synthesized by DNA polymerases.

The entire transcription process should involve the following essential steps:

1. Binding of polymerases to the initiation site. The DNA sequence which signals the initiation of transcription is called the promoter. Prokaryotic polymerases can recognize the promoter and bind to it directly, but eukaryotic polymerases have to rely on other proteins called transcription factors.
2. Unwinding (melting) of the DNA double helix. The enzyme which can unwind the double helix is called helicase. Prokaryotic polymerases have the helicase activity, but eukaryotic polymerases do not. Unwinding of eukaryotic DNA is carried out by a specific transcription factor.
3. Synthesis of RNA based on the sequence of the DNA template strand. RNA polymerases use nucleoside triphosphates (NTPs) to construct a RNA strand.
4. Termination of synthesis. Prokaryotes and eukaryotes use different signals to terminate transcription. [Note: the "stop" codon in the genetic code is a signal for the end of peptide synthesis, *not* the end of transcription.]

Transcription in eukaryotes is much more complicated than in prokaryotes, partly because eukaryotic DNA is associated with histones, which could hinder the access of polymerases to the promotor.

RNA POLYMERASES

The Function of RNA Polymerases

Both RNA and DNA polymerases can add nucleotides to an existing strand, extending its length. However, there is a major difference between the two classes of enzymes: RNA polymerases can initiate a new strand but DNA polymerases cannot. Therefore, during DNA replication, an oligonucleotide (called primer) should first be synthesized by a different enzyme.

The chemical reaction catalyzed by RNA polymerases is shown in figure. The nucleotides used to extend a growing RNA chain are ribonucleoside triphosphates (NTPs). Two phosphate groups are released as pyrophosphate (PP_i) during the reaction. Strand growth is always in the 5' to 3' direction. The first nucleotide at the 5' end retains its triphosphate group.

The Vertical Line Represents the pentose and the slanting line denotes the phosphodiester bond. Bases are designated as N_1, N_2, etc.

Classes of RNA Polymerases

E. Coli

An *E. coli* RNA polymerase is composed of five subunits: two a subunits, and one for each β, β′, and σ subunit. β (151 kD) and β′ (156 kD) are significantly larger than a (37 kD). Several different forms of σ subunits have been identified, with molecular weights ranging from 28 kD to 70 kD. The σ subunit is also known as the σ factor. It plays an important role in recognizing the transcriptional initiation site, and also possesses the helicase activity to unwind the DNA double helix. Nucleotide synthesis is carried out by other four subunits, which together are called the core polymerase. The term "holoenzyme" refers to a complete and fully functional enzyme. In this case, the holoenzyme includes the core polymerase and the σ factor.

Eukaryotes

There are three classes of eukaryotic RNA polymerases: I, II and III, each comprising two large subunits and 12-15 smaller subunits.The two large subunits are homologous to the *E. coli* β and β′ subunits. Two smaller subunits are similar to the *E. coli* α subunit. However, the eukaryotic RNA polymerase does not contain any subunit similar to the *E. coli* α factor. Therefore, in eukaryotes, transcriptional initiation should be mediated by other proteins.

RNA polymerase II is involved in the transcription of all protein genes and most snRNA genes. It is undoubtedly the most important among the three classes of RNA polymerases. The other two classes transcribe only RNA genes. RNA polymerase I is located in the nucleolus, transcribing rRNA genes except 5S rRNA. RNA polymerase III is located outside the nucleolus, transcribing 5S rRNA, tRNA, U6 snRNA and some small RNA genes.

GENE'S REGULATORY ELEMENTS

A gene consists of a transcriptional region and a regulatory region. The transcriptional region is the part of DNA to be transcribed into a primary transcript (an RNA molecule complementary to the transcriptional region). The regulatory region can be divided into cis-regulatory (or cis-acting) elements and trans-regulatory (or trans-acting) elements. The cis-regulatory elements are the binding sites of transcription factors which are the proteins that, upon binding with cis-regulatory elements, can affect (either enhance or repress) transcription. The trans-regulatory elements are the DNA sequences that encode transcription factors.

The cis-acting elements may be divided into the following four types:

PROMOTER

Promoter is the DNA region where the transcription initiation takes place. In prokaryotes, the sequence of a promoter is recognized by the Sigma (σ) factor of the RNA polymerase. In eukaryotes, it is recognized by specific transcription factors.

E. coli

E. coli has five sigma factors:

- Sigma 70: Regulate expression of most genes.
- Sigma 32: Regulate expression of heat shock proteins.
- Sigma 28: Regulate expression of flagellar operon (involved in cell motion).
- Sigma 38: Regulate gene expression against external stresses.
- Sigma 54: Regulate gene expression for nitrogen metabolism.

Eukaryotes

In eukaryotes, there is a significant difference between the transcription of protein genes and RNA genes.

The most common promoter element in eukaryotic protein genes is the TATA box, located at -35 to -20. Its consensus sequence, TATAAA, is quite similar to the -10 region of the Sigma 70 recognition site. Another promoter element is called the initiator (Inr). It has the consensus sequence PyPyAN(T/A)PyPy, where Py denotes pyrimidine (C or T), N = any, and (T/A) means T or A. The base A at the third position is located at +1 (the transcriptional start site).

TATA box and initiator are the core promoter elements. There are other elements often located within 200 bp of the transcriptional start site, such as CAAT box and GC box which may be referred to as promoter-proximal elements.

The protein which interacts with the initiator and TATA box is known as the TATA-box binding protein (TBP), because the TATA box was discovered earlier than the initiator. TBP recognizes not only the core promoter of protein genes, but also RNA promoters. It is α subunit of the general transcription factor TFIID. In eukaryotes, transcription requires several different general transcription factors and, in most cases, the regulatory transcription factors.

Example: The promoter region of the *IL-2* gene - contains TATA box and the binding sites of transcription factors: NFAT, Oct-1, NF-kB and AP-1.

Enhancer

Enhancer is the DNA element that, upon binding with transcription factors (activators), can enhance transcription. It may be located either upstream or downstream of the transcriptional initiation site. However, most of them are located upstream. In prokaryotes, enhancers are quite close to the promoter, but eukaryotic enhancers could be far from the promoter.

(a) *E. coli glnA* gene

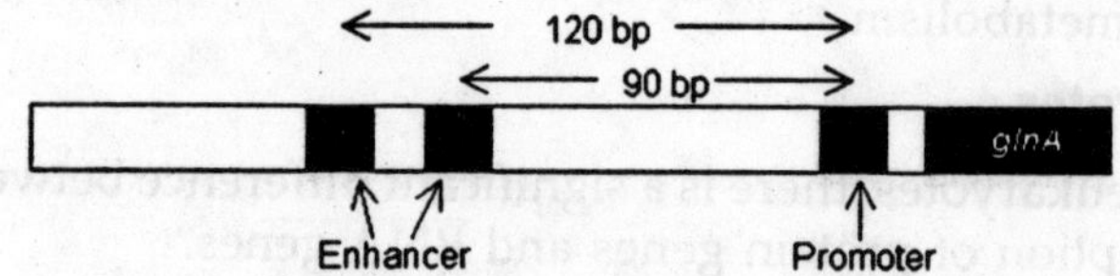

(b) Yeast *GAL1* and *GAL10* genes

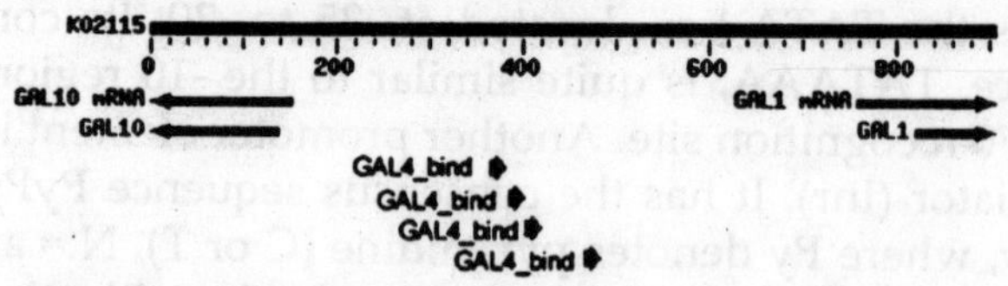

(c) Human b globin gene cluster

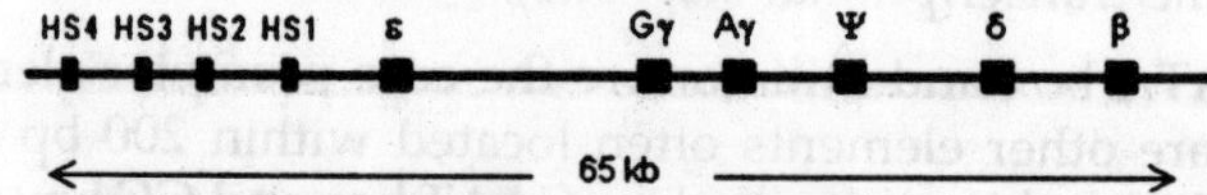

Fig. Examples of Enhancers.

(a) *E. coli glnA* Gene

The enhancer of the *glnA* gene is located about 120 bp from the start site, containing two binding sites of the transcription factor nitrogen regulatory protein C (NTRC).

(b) Yeast *GAL1* and *GAL10* Genes

Both *GAL1* and *GAL10* genes are regulated by the same enhancer between them. The two genes are transcribed along opposite directions. Therefore, the enhancer is located upstream of both transcriptional start sites. It is also called upstream activating sequences (UAS), which contains four binding sites of the transcription factor GAL4.

Human β Globin Gene Cluster

The human β globin gene cluster is controlled by an enhancer region comprising HS1 to HS4, which contain the binding sites of GATA-1, NF-E2, AP-1 and other transcriptional activators (illustration from Globin Gene Server). This region is known as the locus control region (LCR), which regulates the expression of all five genes (ε, Gγ, Aγ, δ and β), even though the distance between HS4 and the β gene is as far as 60 kb.

In embryonic DNA, the ε gene is preferentially expressed. In fetal DNA, Gγ and Aγ are much more strongly expressed than the other genes. In adult DNA, expression is switched to mainly the β gene, while the δ gene is weakly expressed.

Table Examples of Transcriptional Activators in Mammals.

Transcription Activator	Consensus Sequence	Transcription Activator	Consensus Sequence
AP-1	TGAGTCA	p53	PuGPuCATGPyCPy
AP-2	CCC(A/C)N(C/G)$_3$	NF-kB	GGGPuNTPyPyCC
Oct-1	ATGCAAAT	NFAT	GGAGAPu
GATA-1	(A/T)GATAPu	NF-E2	TGACTCAG

*Pu = Purine (A or G); Py = Pyrimidine (C or T); N = any

Silencer

Silencer is the DNA element that, upon binding with transcription factors (repressors), can repress transcription. In prokaryotes, silencers are known as operators, found in many genes such as *lac* operon and *trp* operon. In eukaryotes, the following genes have been demonstrated to contain silencers

In a few cases, a DNA element may act either as an enhancer or a silencer, depending on the binding protein. For example, certain genes contain an element called E box (consensus CACGTG) which can bind either Max/Myc dimer or Max/Mad dimer. The Max/Myc dimer activates transcription, whereas the Max/Mad dimer suppresses transcription of these genes.

Response Elements

Response elements are the recognition sites of certain transcription factors. Most of them are located within 1 kb from the transcriptional start site.

Table. Eukaryotic Response Elements

Response Element	Transcription Factor	Consensus Sequence
CRE	CREB	TGACGTCA
ERE	Estrogen receptor	AGGTCANNNTGACCT
GRE	Glucocorticoid receptor	AGAACANNNTGTTCT
HSE	Heat shock factor	GAANNTTCNNGAA
SRE	Serum response factor	$CC(A/T)_6GG$

*$(A/T)_6$ means six A or T; N = any.

The cAMP response element (CRE) interacts with CREB (CRE-binding protein), which is regulated by cAMP.

Estrogen response element (ERE) and glucocorticoid response element (GRE) are the recognition sites of estrogen receptor and glucocorticoid receptor, respectively. Note that hormones are *not* transcription factors, but many of their receptors are.

Heat shock response element (HSE) is present in heat shock protein genes. In response to external stress (e.g. high temperature), the heat shock factor (HSF) will interact with HSE, stimulating expression of heat shock proteins.

Serum response element (SRE) binds to serum response factor (SRF), which can be activated by many growth factors in serum. The Fos subunit of AP-1 is encoded by a gene containing SRE. Fos is known to play an important role in cell cycle progression.

Its ligand binding domain binds to the steroid and the DNA binding domain interacts with DNA to regulate gene transcription.

The transcription region consists of exons and introns. The regulatory elements include promoter, response element, enhancer and silencer (not shown). Downstream refers to the direction of transcription and upstream is opposite to the transcription direction. The numbering of base pairs in the promoter region is as follows. The number increases along the direction of transcription, with "+1" assigned for the initiation site. There is no "0" position. The base pair just upstream of +1 is numbered "-1", not "0".

TRANSCRIPTION MECHANISMS IN PROKARYOTES

In prokaryotes, binding of the polymerase's s factor to promoter can catalyze unwinding of the DNA double helix. The most important s factor is Sigma 70, whose structure has been determined by x-ray crystallography.

Note that the promoter is rich in A and T. The AT pair involves two hydrogen bonds whereas the CG pair involves three hydrogen bonds. Therefore, AT pairs are easier to separate. The DNA replication origin is also rich in A and T.

After the DNA strands are separated at the promoter region, the core polymerase ($\alpha\alpha\beta\beta'$) can then start to synthesize RNA based on the sequence of the DNA template strand. Since the role of the s factor is mainly to initiate transcription, it will be released after about 10 ribonucleotides have been polymerized.

Elongation of the RNA strand continues until the core polymerase reaches the termination site (more info).

Regulation by Transcription Factors

Regulation of the *lac* operon transcription by

Catabolite Activator Protein (CAP) and *lac* Repressor

Activation of the *glnA* transcription by

Nitrogen Regulatory Protein C (NTRC)

(Example of DNA looping)

TRANSCRIPTIONAL TERMINATION IN PROKARYOTES

In prokaryotes, the transcription is terminated by two major mechanisms: Rho-independent (intrinsic) and Rho-dependent.

The Rho-independent termination signal is a stretch of 30-40 bp sequence, consisting of many GC residues followed by a series of T ("U" in the transcribed RNA). The resulting RNA transcript will form a stem-loop structure to terminate transcription.

Rho-dependent Mechanism

Rho is a ~ 50 kD protein, involved in bout half of *E. coli* transcriptional terminations. It has been well established that six Rho proteins form a hexamer to terminate transcription, but the precise mechanism is not clear. Experiments suggest that two components are essential: (i) the upstream Rho loading site and (ii) the downstream termination site. The Rho hexamer first binds to the RNA transcript at the upstream site which is 70-80 nucleotides long and rich in C residues. Upon binding, the Rho hexamer moves along the RNA in the 3' direction. If movement of the polymerase is slow, the Rho hexamer will catch up and terminate the transcription at the downstream termination site. Rho has ATPase activity which can induce release of the polymerase from DNA.

REGULATION OF LAC OPERON TRANSCRIPTION

Transcriptional Activation by CAP

In the absence of any transcriptional activator, binding of the polymerase to the promoter of the *lac* operon is weak. The binding site of the catabolite activator protein (CAP) is located just upstream of the promoter. In such case, the activator can enhance transcription by direct contact with the polymerase, making it bind more efficiently to the promoter.

Transcriptional Inhibition by *Lac* Repressor

The silencer (also known as operator) of the *lac* operon is located at the transcriptional start site. When the *lac* repressor binds to this region, it will block the movement of the polymerase, thereby inhibiting transcription.

ACTIVATION OF GLNA TRANSCRIPTION BY NTRC

Unlike the *lac* operon, the enhancer of the *glnA* gene is a little far from the promoter so that the bound activator does not immediately contact the polymerase. Nitrogen Regulatory Protein C (NTRC) can induce DNA looping to bring the activator in contact with the polymerase.

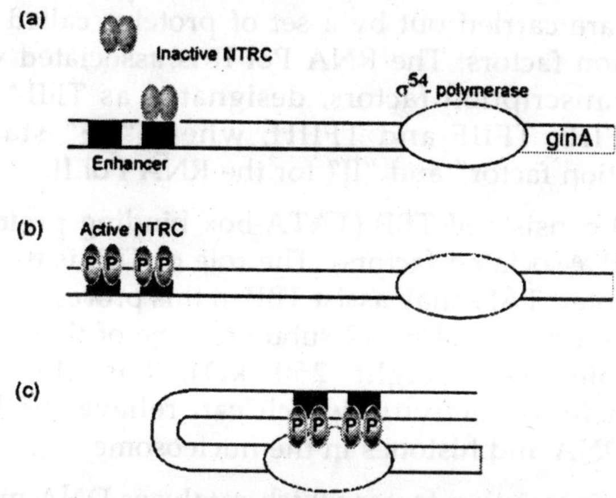

Fig. The Mechanism of Transcriptional Activation by NTRC.

a. The *glnA* gene is transcribed by the Sigma-54-containing polymerase which alone cannot initiate transcription. The unphosphorylated NTRC dimers can bind only one site at the enhancer, still insufficient to stimulate transcription.
b. The phosphorylated NTRC dimers can bind both sites of the enhancer.
c. Their binding induces DNA looping. Contact between the activator and the polymerase stabilizes the

interaction between the polymerase and DNA, thereby initiating transcription.

TRANSCRIPTION MECHANISMS IN EUKARYOTES

In Eukaryotes, there are three classes of RNA polymerases: I, II and III. This section will focus on the RNA polymerase II (Pol II), which is involved in the transcription of all protein genes. Transcription by RNA Pol I and Pol III is discussed.

Initiation

RNA Pol II does not contain α subunit similar to the prokaryotic s factor, which can recognize the promoter and unwind the DNA double helix. In eukaryotes, these two functions are carried out by a set of proteins called general transcription factors. The RNA Pol II is associated with six general transcription factors, designated as TFIIA, TFIIB, TFIID, TFIIE, TFIIF and TFIIH, where "TF" stands for "transcription factor" and "II" for the RNA Pol II.

TFIID consists of TBP (TATA-box binding protein) and TAFs (TBP associated factors). The role of TBP is to bind the core promoter. TAFs may assist TBP in this process. In human cells, TAFs are formed by 12 subunits. One of them, TAF250 (with molecular weight 250 kD), has the histone acetyltransferase activity, which can relieve the binding between DNA and histones in the nucleosome.

The transcription factor which catalyzes DNA melting is TFIIH. However, before TFIIH can unwind DNA, the RNA Pol II and at least five general transcription factors (TFIIA is not absolutely necessary) have to form a pre-initiation complex (PIC).

ASSEMBLY OF THE PRE-INITIATION COMPLEX (PIC)

TBP first binds to the promoter and then recruits TFIIB to join TFIID (and TFIIA if present). Before entering PIC, RNA Pol II and TFIIF are bound together, which are recruited by TFIIB. Finally, RNA Pol II recruits TFIIE, which further recruits TFIIH to complete the PIC assembly.

HISTONE ACETYLATION, DNA METHYLATION AND CHROMATIN REMODELING

HISTONE ACETYLATION

Acetylation of the lysine residues at the N terminus of histone proteins removes positive charges, thereby reducing the affinity between histones and DNA. This makes RNA polymerase and transcription factors easier to access the promoter region. Therefore, in most cases, histone acetylation enhances transcription while histone deacetylation represses transcription.

Histone acetylation is catalyzed by histone acetyltransferases (HATs) and histone deacetylation is catalyzed by histone deacetylases (denoted by HDs or HDACs). Several different forms of HATs and HDs have been identified. Among them, CBP/p300 is probably the most important, since it can interact with numerous transcription regulators.

CREB BINDING PROTEIN (CBP) AND P300

CBP and p300 are highly related histone acetyltransferases (HATs). The following figure shows the domain structure of p300.

"Bromo" represents Bromodomain which specifically binds to acetylated lysine. "CH-1, CH-2 and CH-3" are cysteine/histidine-rich regions.

TRANSCRIPTIONAL REGULATION BY HISTONE ACETYLATION AND DEACETYLATION

p300/CBP can interact with a variety of transcriptional regulators such as pCAF (a histone acetyltransferase) and TBP (which recognizes the promoter). During transcription, they are all assembled at the promoter region. Histone acetylation by p300/CBP. pCAF and $TAF_{II}250$ facilitates the transcription.

The repressor Mad/Max dimer interacts with SIN3 (or N-CoR, SMRT, etc.) which recruits histone deacetylase (HDs) to repress transcription.

REGULATION OF TRANSCRIPTION FACTORS

The action of transcription factors can either stimulate or inhibit gene expression. They must be regulated so that a gene is expressed only when its product is needed. The following are a few examples.

Regulation of lac Repressor

The *lac* operon of *E. coli* consists of three genes: *lacZ, lacY,* and *lacA,* encoding b-galactosidase, lactose permease and thiogalactoside transacetylase, respectively. Lactose permease is located on the cell membrane, capable of pumping lactose into the cell. b-galactosidase can convert lactose into glucose and galactose. Thiogalactoside transacetylase is responsible for degrading small molecules.

In the absence of lactose, transcription of the *lac* operon is inhibited by the *lac* repressor. The lactose can bind to the *lac* repressor, preventing it from interacting with its DNA binding site. Hence, in a medium containing lactose, the *lac* operon is quickly transcribed, producing the enzymes to generate glucose, which is the major energy source for *E. coli.*

Regulation of lambda Repressor (cI)

The life cycle of l phages is controlled by cI and Cro proteins. The l phage will remain in the lysogenic state if cI proteins predominate, but will be transformed into the lytic cycle if Cro proteins predominate. Transcription of the two proteins is regulated by the cI protein (l repressor) itself.

Figure Regulation of the expression of cI and Cro proteins by cI. The cI dimer may bind to any of three operators, O_R1, O_R2, and O_R3, in the order $O_R1 > O_R2 > O_R3$. Binding of a cI dimer to O_R1 enhances binding of a second cI dimer to O_R2. Thus, O_R1 and O_R2 are almost always simultaneously occupied by cI. However, this does not increase the affinity between cI and O_R3, which will be occupied only when the cI concentration is high.

a. In the absence of cI proteins, the *cro* gene may be transcribed.

b. In the presence of cI proteins, only the *cI* gene may be transcribed.

c. At high concentration of cI, transcriptions of both genes are repressed.

When the host DNA is damaged (e.g., under UV irradiation), the cI protein may be cleaved by certain protease promoted by the RecA protein. Cleaved cI proteins cannot bind to the operators. Thus, the Cro proteins can be produced to transform the l phage into the lytic cycle.

Regulation of NF-kB

Nuclear factor-kB (NF-kB) is a ubiquitous transcription factor that controls the expression of genes involved in immune responses, apoptosis, and cell cycle. Incorrect regulation of NF-kB may cause inflammatory and autoimmune diseases, viral infection and cancer. Five mammalian NF-kB family members have been identified:

- NF-kB1 (also called p50)
- NF-kB2 (also named p52)
- RelA (also known as p65)
- RelB
- c-Rel

They all share a highly conserved Rel homology domain, responsible for their dimerization and binding to DNA and IkB (inhibitor of NF-kB). The transcription factor NF-kB works only when two members form a dimer. The most abundant activated form consists of a p50 or p52 subunit and a p65 subunit.

NF-kB can be activated by a variety of stimuli, including cytokines (such as TNF-α and IL-1), T and B cell mitogens, viral proteins, and stress inducers (such as reactive oxygen species or UV radiation). In the cytoplasm, NF-kB is inhibited by IkB. Upstream activating signal (e.g., binding of TNF-α to its receptor) may cause phosphorylation of IkB by IKK (IkB kinase). This triggers the degradation of IkB through the ubiquitin system, where the target molecule is masked by a chain of ubiquitins for degradation by the 26S protesome. The free NF-kB can then translocate to the nucleus and activate transcription.

P53

p53 is a tumor suppressor protein, also known as "Guardian of the Genome". It plays an important role in cell cycle control and apoptosis. Defective p53 could allow abnormal cells to proliferate, resulting in cancer. As many as 50per cent of all human tumors contain p53 mutants.

In normal cells, the p53 protein level is low. DNA damage and other stress signals may trigger the increase of p53 proteins, which have three major functions: growth arrest, DNA repair and apoptosis (cell death). The growth arrest stops the progression of cell cycle, preventing replication of damaged DNA. During the growth arrest, p53 may activate the transcription of proteins involved in DNA repair. Apoptosis is the "last resort" to avoid proliferation of cells containing abnormal DNA.

The cellular concentration of p53 must be tightly regulated. While it can suppress tumors, high level of p53 may accelerate the aging process by excessive apoptosis. The major regulator of p53 is Mdm2, which can trigger the degradation of p53 by the ubiquitin system.

Target Genes

p53 is a transcriptional activator, regulating the expression of *Mdm2* (for its own regulation) and the genes involved in growth arrest, DNA repair and apoptosis. Some important examples are listed below.

1. Growth arrest: *p21, Gadd45, and 14-3-3s.*
2. DNA repair: *p53R2.*
3. Apoptosis: *Bax, Apaf-1, PUMA and NoxA.*

Regulation of p53

As mentioned above, p53 is mainly regulated by Mdm2. The regulation mechanism is illustrated in the following figure.

a. Expression of Mdm2 is activated by p53.

b. Binding of p53 by Mdm2 can trigger the degradation of p53 via the ubiquitin system.

c. Phosphorylation of p53 at Ser15, Thr18 or Ser20 will disrupt its binding with Mdm2. In normal cells, these three residues are not phosphorylated, and p53 is maintained at low level by Mdm2.

d. DNA damage may activate protein kinase (such as ATM, DNA-PK, or CHK2) to phosphorylate p53 at one of these three residues, thereby increasing p53 level. Since Mdm2 expression is activated by p53, the increase of p53 also increases Mdm2, but they have no effect while p53 is phosphorylated. After the DNA damage is repaired, the ATM kinase is no longer active. p53 will be quickly dephosphorylated and destroyed by the accumulated Mdm2.

Roles of p53

The roles of p53 in growth arrest and apoptosis are illustrated in figure. p53 is also directly involved in DNA repair. One of its transcriptional target gene, *p53R2*, encodes ribonucleotide reductase, which is important for both DNA replication and repair. p53 also interacts directly with AP endonuclease and DNA polymerase which are involved in base excision repair.

a. The cell cycle progression into the S phase requires the enzyme Cdk2, which can be inhibited by p21. The progression into the M phase requires Cdc2 which can be inhibited by p21, GADD45 or 14-3-3s. p53 regulates the expression of these inhibitory proteins to induce growth arrest.

b. Apoptosis can be induced by the binding of Caspase 9 to cytochrome·c and Apaf1. p53 may activate the expression of Apaf1 and Bax. The latter can then stimulate the release of cytochrome c from mitochondria.

TRANSCRIPTION OF RNA GENES

The products of RNA genes include rRNA, tRNA and small RNA molecules. In the mammalian genome, the three rRNA genes, 28S, 18S, and 5.8S, are clustered as a pre-rRNA gene. which is transcribed by RNA polymerase I. tRNA, 5S rRNA and the U6 snRNA genes are transcribed by RNA polymerase III. Most snRNA genes are transcribed by RNA polymerase II using the same mechanism as transcribing protein genes.

TRANSCRIPTION BY RNA POLYMERASE I

RNA polymerase I is devoted to the synthesis of pre-rRNA. Like RNA polymerase II, a pre-initiation complex (PIC) has to be formed before RNA polymerase I can exert its function. The order of the PIC assembly is illustrated in figure below.

Figure. The PIC assembly for RNA polymerase I. The regulatory region of pre-rRNA gene contains a core element and an upstream control element (UCE). Binding of two upstream binding factors (UBF) to both elements may induce DNA looping, and subsequently recruiting TATA-binding protein (TBP) and TBP-associated factors (TAF_I). Finally, RNA polymerase I joins the complex and completes the assembly process.

TRANSCRIPTION BY RNA POLYMERASE III

Transcription of tRNA gene

The regulatory region of tRNA gene contains A box and B box located inside the transcription unit. The PIC assembly begins with the binding of TFIIIC to both elements.

Transcription of 5S-rRNA gene

The regulatory region of 5S-rRNA gene contains a C box, also located inside the transcription unit. The PIC assembly begins with the binding of TFIIIA to the C box.

GENOMIC REPLICATION OF VIRUSES

CLASS I VIRUSES

Type I: dsDNA

Replication Strategy of dsDNA Viruses

Steps in Replication

1. Primary transcription by host enzymes
2. Translation of early (=regulatory) proteins
3. Viral genomic DNA replication (usually by host enzymes)
4. Late transcription (usually mediated by viral proteins)
5. Synthesis of late (=structural) proteins
6. Assembly of structural protein and DNA into virions

dsDNA viruses include:

- Viruses infecting Eubacteria and Archaea, or phages (eg. the *Podoviridae,* phages of *E coli*);
- Viruses of higher animals (*Pox-, Herpes-, Adeno-, Papovaviridae*);
- Viruses of insects (*Baculo-* and *Irido- and Polydnaviridae*);
- And viruses of eukaryotic algae (*Phycodnaviridae*);
- Viruses of fungi

These range in size from:

- 5-8 kb (*Papovaviridae*)
- Through 30-40 kb (lambda phage and *Adenoviridae*)
- To about 200 kb (*Baculoviridae*)
- To over 300 kb (*Herpes- and Pox-* and *Irido-* and *Phycodnaviridae*).

By contrast, the smallest prokaryote genome size is about 600 kb (*E coli* = 4000 kb)

They may have:

- Circular genomes (*Papova-* and *Baculo-* and *Polydnaviridae*);

- Linear genomes (*Adeno-* and *Herpesviridae,* some phages);
- Have circularly permuted linear genomes (phage T4, some *Iridoviridae*);
- Or linear genomes with covalently closed ends (*Pox-* and *Phycodnaviridae*).

All viruses except *Polydnaviridae* have single-component genomes; the latter have multiple components ranging in size from 2 - 20 kb, and the number which constitute an individual genome is not known.

Replication of the viruses is in all cases by the semi-conservative method favoured by cellular genomes; however, smaller circular genomes (eg. *Papovaviridae*) replicate by means of bidirectional replication forks from a single origin, like some plasmids. Among the viruses of Eukarya, replication mainly occurs in the nucleus, using cellular enzymes such as polymerases, methylases, etc. However,the replication of Poxviruses, some Baculoviruses (granulosis group), and some of the replication of iridoviruses, takes place in virus-specified "inclusion bodies" in the cytoplasm, using viral-coded enzymes, most important of which are DNA-dependent DNA polymerases.

Herpesviruses

Name comes from the Greek *'Herpein'* - 'to creep' = chronic/latent/recurrent infections. Epidemiology of the common Herpesvirus infections puzzled clinicians for many years. In 1950, Burnet and Buddingh showed that HSV could become latent after a primary infection, becoming reactivated after later provocation. Weller (1954) isolated VZV from chicken pox and zoster, indicating the same causal agent. ~100 Herpesviruses have been isolated, at least one for most animal species which has been looked at. To date, there are 8 known human Herpesviruses.

The family is divided into 3 Sub-families:

Alphaherpesvirinae:	
Simplexvirus	human herpesvirus 1, 2 (HSV-1, HSV-2)
Varicellovirus	human herpesvirus 3 (VZV)

Betaherpesvirinae:	
Cytomegalovirus	human herpesvirus 5 (CMV)
Muromegalovirus	mouse cytomegalovirus 1
Roseolovirus	human herpesvirus 6, 7 (HHV-6, HHV-7)
Gammaherpesvirinae:	
Lymphocryptovirus	human herpesvirus 4 (EBV)
Rhadinovirus	human herpesvirus 8 (HHV-8)

Large (genomes up to 235kbp DNA) and complex viruses ~35 virion proteins. All encode a variety of enzymes involved in nucleic acid metabolism, DNA synthesis and protein processing (protein kinase). The Herpesviruses are widely separated in terms of genomic sequence and proteins, but all are similar in terms of virion structure and genome organization:

Size	:	180-200nm
Envelope	:	Present; associated glycoproteins.
Tegument	:	Protein-filled region between capsid and envelope.
Capsid	:	Icosahedral, 95-105nm diameter; 162 hexagonal capsomers.
Core	:	Toroidal (DNA around protein), ~75nm diameter.
Genome	:	Linear, d/s DNA, 130-230kbp
Replication	:	Nuclear.
Assembly	:	Nuclear.
Common Antigens	:	None!

Structure

The structure of the herpesvirus particle is very complex. The core consists of a toroidal shape with the large DNA genome would around a proteinaceous core. The complex capsid surrounds the core. Outside the capsid is the tegument, a protein-filled region which appears amorphous in electron micrographs. On the outside of the particle is the envelope, which contains numerous glycoproteins. To view an electron micrograph of negatively-stained herpesvirus particles.

N.B. All herpesviruses are almost indistinguishable in electron micrographs.

Genome

All herpesvirus genomes have a unique long (UL) and a unique short (US) region, bounded by inverted repeats. The repeats allow rearrangements of the unique regions and Herpesvirus genomes exist as a mixture of 4 isomers. Herpesvirus genomes also contain multiple repeated sequences and depending on the number of these, genome size of various isolates of a particular virus can vary by up to 10kbp.

The prototype member of the family is Herpes Simplex Virus (HSV): ~160kbp - complete sequence is known. Prior to this, the HSV genome had been mapped extensively by a large number of t.s. mutants. There are 2 antigenic types, HSV-1 and HSV-2 which share antigenic cross-reactivity but different neutralization patterns and tend to produce different clinical symptoms. Man is believed to be the natural host for HSV, but the virus is also capable of infecting various animals, including rodents (good animal models). Human infection is virtually universal - most adults are seropositive.

Replication

The HSV envelope contains at least 9 glycoproteins. Most of these appear to be individually dispensable for infectivity. Interaction of glycoprotein(s) with cellular receptors (not known, but probably different receptors on different cell types - a characteristic of the larger, more complex DNA viruses, and probably common cell surface molecules rather than specific proteins) results directly in fusion of the envelope with the cell membrane. Endocytosis is not absolutely required, but may occur (alternative route for penetration).

Fusion deposits the capsid in the cytoplasm, where it migrates to the nucleus. The core enters via a nuclear pore where the genome is circularized.

Transcription of the large, complex genome is sequentially regulated in a cascade fashion. ~50 mRNAs are produced by host cell RNA polymerase II.

Three distinct classes of mRNAs are made:

Alpha: immediate early (IE) mRNAs	5 trans-acting regulators of virus transcription
Beta: (delayed) Early mRNAs	Further non-structural regulatory proteins and minor structural proteins
Gamma: Late mRNAs	Major structural proteins

Gene expression is co-ordinately regulated:

- If translation is blocked shortly after infection, IE mRNAs accumulate in the nucleus, but no other virus mRNAs are transcribed.
- Synthesis of early gene products turns off IE products and initiates genome replication.
- Some of the late structural proteins (gamma-1) are produced independently of genome replication, others (gamma-2) are only produced after replication.

Both IE and E proteins are required for genome replication. A virus-encoded DNA-dependent DNA polymerase and DNA-binding protein are involved in replication, together with a number of enzymes (e.g. thymidine kinase) which alter cellular biochemistry. In addition, cellular proteins are required for genome replication, therefore HSV replication occurs in the nucleus (c.f. Poxviruses).

Viral DNA replication is the target for a number of successful anti-Herpesvirus drugs (e.g. acyclovir, gancyclovir, etc). The pattern of replication is complex, involving at least 3 potential origins of replication, and resulting in the formation of high molecular weight DNA concatemers.

Virus particles (core plus capsid) assemble in the nucleus - genomic concatemers are cleaved and packaged into pre-assembled capsids.

The envelope is acquired from the inner lamella of the nuclear membrane, and particles accumulate in the space within the inner and outer lamellae. How these particles are transported to the cell surface is not clear and may or may not involve the golgi apparatus. Mutations in certain envelope glycoproteins interfere with cytoplasmic transport. Any remaining virus particles are released when the cell lyses (~24h after infection).

HSV infection appears to be a 'wasteful' process, only ~25per cent of viral DNA/protein produced is incorporated into virions. The rest accumulates in the cell, which eventually dies. This process produces characteristic nuclear inclusion bodies.

Pathogenesis

Herpes Simplex (HSV)

Primary infection occurs through a break in the mucus membranes of the mouth or throat, via the eye or genitals or directly via minor abrasions in the skin. because of the universal distribution of the virus, most individuals are infected by 1-2 years of age; initial infection is usually asymptomatic, although there may be minor local vesicular lesions. Local multiplication ensues, followed by viraemia and systemic infection. There then follows life-long latent infection with periodic reactivation.

During primary infection, the virus enters peripheral sensory nerves and migrates along axons to sensory nerve ganglia in the CNS - allows virus to escape immune response! During latent infection of nerve cells, viral DNA is maintained as an episome (not integrated) with limited expression of specific virus genes required for the maintenance of latency - true latency.

The delicate balance of latency may be upset by various disturbances, physical (injury, U.V, hormones, etc) or psychological (stress, emotional upset - perhaps affecting immune system/hormonal balance).

Reactivation of latent virus leads to recurrent disease - virus travels back down sensory nerves to surface of body and replicates, causing tissue damage.

Poxviruses

Introduction

The largest and most complex viruses known! Have been known about for centuries - the characteristic "pocks" produced by variola virus (Smallpox) giving their name to all forms of infectious disease "a dose of the pox". Smallpox first appeared in China and the Far East at least 2000 years ago.

Taxonomy

Subfamily:	**Genus:**	**Members:**
	Avipoxvirus	fowlpox virus
	Capripoxvirus	sheeppox virus
	Leporipoxvirus	myxoma virus
Chordopoxvirinae	Molluscipoxvirus contagiosum	Molluscum
	Orthopoxvirus	vaccinia virus
	Parapoxvirus	orf virus
	Suipoxvirus	swinepox virus
	Yatapoxvirus	Yaba monkey tumor virus
Entomopoxvirinae	Entomopoxvirus A melolontha	Melolontha
entomopoxvirus		
	Entomopoxvirus B entomopoxvirus	Amsacta moorei
	Entomopoxvirus C	Chironomus luridus entomopoxvirus

Morphology

Oval or "brick-shaped" particles 200-400nm long - can be visualized by the best light microscopes (just). The external surface is ridged in parallel rows, sometimes arranged helically. The particles are extremely complex, containing many proteins (more than 100) and detailed structure is not known.

The extracellular forms contain 2 membranes (EEV - extracellular enveloped virions), intracellular particles only have an inner membrane (IMV - intracellular mature virions).

Thin sections in E.M. reveal that the outer surface is composed of lipid and protein which surrounds the core, which is biconcave (dumbbell-shaped), with 2 "lateral bodies" (function unknown). The core is composed of a tightly compressed nucleoprotein. Antigenically, poxviruses are very complex, inducing both specific and cross-reacting antibodies - hence ability to vaccinate against one disease with another virus (below). There are at least 10 enzymes present in the particle, mostly concerned with nucleic acid metabolism/ genome replication.

Genome

Linear, d/s DNA of 130-300kbp. Ends of genome consist of a terminal hairpin loop (no free ends) with several tandem (i.e. direct) repeat sequences (this arrangement is found at the ends of chromosomes from a number of different organisms). The ends of the genome form direct repeats called inverted terminal repeats (ITRs). Several poxvirus genomes have been sequenced. Most of the essential genes are located in the central part of the genome, while non-essential (in tissue culture) genes are located at the ends. There are ~200 genes in the genome.

Replication

Occurs in the cytoplasm - the virus is sufficiently complex to have acquired all the functions necessary for genome replication (c.f. Herpesviruses). There is some contribution from the cell but it is not clear what this is - poxvirus gene expression and genome replication occur in enucleated cells, but maturation is blocked.

Receptors are not known, but probably >1 on different cell types. For Vaccinia, one of these is probably the EGF receptor (epidermal growth factor).

Penetration is complex and may also involve >1 mechanism.

Uncoating occurs in two stages, removal of the outer membrane as the particle enters the cell and in the cytoplasm, the particle (minus its outer membrane) is further uncoated and the core passes into the cytoplasm.

Gene expression is carried out (exclusively?) by viral enzymes associated with the core and is divided into 2 phases:

Early genes: ~50per cent genome, expressed before genome replication

Late genes: expressed after genome replication; late promoters are dependent on DNA replication for activity.

Poxvirus gene expression has been studied in detail because of the interest in the use of Vaccinia virus as a vector for expression of heterologous genes.

Genome replication is believed to involve self-priming, leading to the formation of high m.w. concatemers (isolated from infected cells) which are subsequently cleaved and repaired to make virus genomes. The many virus-encoded enzymes involved in replication (e.g. thymidine kinase - *tk*) offer potential targets for chemotheraputic agents.

Assembly occurs in the cytoskeleton; the events involved in putting together such a complex particle are not understood, but probably involve interactions with the cytoskeleton (e.g. actin-binding proteins). Inclusions are formed in the cytoplasm which mature into virus particles. Actin 'comet tails' form which shoot IEV through the cytoplasm to the cell surface, and possibly into adjacent cells - these have been timed moving at 3μm/min. This may provide an alternative mechanism for cell to cell spread (c.f. EEV).

Overall, replication of this large, complex virus is rather quick ~12h.

Vaccinia virus shows considerable resistance to the antiviral effects of interferons. One of the early genes of this virus, K3L, encodes a protein which is homologous to eIF-2a which inhibits the action of PKR. In addition, the E3L protein also binds dsRNA and inhibits PKR activation. As well as these, other poxvirus-encoded proteins interfere with the actions of complement, IL-1 and TNFs. Vaccinia infection of cells can confer "protection" from IFN on other viruses, e.g. Picornaviruses.

Type II: ssDNA

ssDNA viruses include organisms infecting:

- Bacteria ("bacteriophages", eg.*Inoviridae, Microviridae*),
- Mammals (*Circoviridae, Parvoviridae* click for image),
- Birds (circovirus-like organisms),
- And plants (*Geminiviridae,* banana bunchy top-like viruses or Nanoviruses).

They can have:

- Linear single-component genomes (*Parvoviridae*),
- Circular single-component genomes (*Microviridae, Inoviridae, Circoviridae,* some *Geminiviridae*),
- Circular two-component genomes (some *Geminiviridae*),
- Or circular multicomponent (>3) genomes (Nanoviruses).

The genomes are all relatively small:

- The *Circoviridae* have genomes of about 3 kb;
- *Parvoviridae* have genomes of 4-5 kb;
- The *Geminiviridae* about 2.7 - 5.4 kb (depending on whether are mono- or bi-component);
- The *Microviridae* (including the famous phiX 174) about 4.5 kb;
- The *Inoviridae* and the Nanoviruses about 5-6 kb.

Replication of all of the viruses requires formation of a "replicative form" (RF) double- stranded DNA intermediate: this is formed soon after infection, almost certainly by the host cell DNA polymerases engaging in "repair" of the ssDNA.

In the case of circular genomes in Eukarya (circoviruses, eg. porcine circovirus; also psittacine beak-and-feather disease agent, chicken infectious anaemia agent; *Geminiviridae;*

BBTRV), these get converted into cccds-(plasmid-like)-DNA in the nucleus, and become associated with nuclear proteins and complexes such as nucleosomes.

Parvoviruses have an interesting strategy for replicating their genomes, which uses internal or self-complementarity of genome ends to get around the problem of how to replicate a linear DNA genome. A virus-specific process is required both to nick RF DNA, and to sequester newly-formed genomic ssDNA into assembling particles: in the case of Parvoviridae the first is done by a NS1 protein (which binds the new 5'-terminus resulting from the nick) and the second by the coat protein; in *Geminiviridae* it appears as if the first is done by a similarly-acting Rep protein, and the second also by the coat protein.

Parvoviruses

Introduction

Parvoviruses are among the smallest, simplest eukaryotic viruses and were only discovered in the 1960's. They are widespread in nature; human Parvovirus infections were only recognised in the 1980's. Essentially, they fall into two groups, defective viruses which are dependent on helper virus for replication and autonomous, replication-competent viruses. In all, >50 parvoviruses have been identified.

The Family Contains 6 Genera Divided Between 2 Subfamilies:

Genus:	Type species:	Host organism:
	Parvovirinae:	
Parvovirus	Mice minute virus	Vertebrates
Erythrovirus	B19 virus	Vertebrates
Dependovirus	Adeno-associated virus 2	Vertebrates
	Densovirinae:	
Densovirus	Junonia coenia densovirus	Invertebrates
Iteravirus	Bombyx mori densovirus	Invertebrates
Contravirus	Aedes aegypti densovirus	Invertebrates

The Dependoviruses were the first to be discovered. Unlike the other 5 genera, these are replication-defective viruses which are entirely dependent helper viruses. The genus consists of 4 types of adeno-associated viruses (AAV1-4) which require Adenoviruses (or Herpesviruses) for helper functions to replicate. The adenovirus genes involved are early (regulatory) rather than late (structural), such as E1A, but it has recently been shown that treatment of cells with U.V, cycloheximide or some carcinogens can replace the requirement for helper virus. Therefore the requirement appears to be for a modification of the cellular environment rather than a specific virus protein - probably affecting transcription of the defective virus genome.

Morphology

Particles are icosahedral, 18-26nm diameter and consist only of protein (50per cent) + DNA (50per cent). There are 3 capsid proteins, VP1-3. VP3 is derived by protease cleavage of VP2. Infectious virions of canine parvovirus contain 60 protein subunits that are predominantly VP-2. The central structural motif of VP-2 has the same topology (an eight-stranded antiparallel beta-barrel) as has been found in many other icosahedral viruses (e.g. picornaviruses).

The capsid confers considerable stability on the virions, which are resistant to inactivation by pH, solvents or high temperatures (1h @ 50°C).

Genome

Linear, non-segmented, s/s DNA, ~5kb. Most of the strands packaged seem to be (-)sense, but AAVs package equal amounts of (+) and (-) strands, and all seem to package at least a proportion of (+)sense strands. The ends of the genome have palindromic sequences of ~115nt which form "hairpins". These structures are essential for the initiation of genome replication.

Replication

The pathogenic human parvovirus B19 is difficult to grow in culture, so comparatively little is known about its biology. The receptor molecule for B19 is tetrahexosoceramide, a glycolipid (erythrocyte P antigen) - explains cell tropism of B19:

Replication occurs in the nucleus and is thought to follow the scheme above. All Parvoviruses are highly dependent on cellular functions for genome replication. The autonomous viruses require the cell to pass through S-phase for replication to occur, but unlike Polyomaviruses (below) cannot turn on host cell DNA synthesis. The defective viruses are even more dependent - require host cell machinery + helper virus for replication.

Expression of the Parvovirus genome is poorly understood. The helper function required by the defective viruses is thought to be involved in transcription of virus genes. Host cell DNA polymerase is necesary for genome replication.

In the absence of helper virus, Dependoviruses can establish a latent infection; in this state, the virus genome is integrated into the host cell DNA - and can be rescued >50 passages later by Adenovirus infection. The rep gene is involved in this process, which is poorly understood. There is evidence for vertical transmission of avian AAV in chickens. Conversely, AAV appears to inhibit cellular transformation by Adenoviruses.

Pathogenesis

Parvoviruses cause infections in a wide variety of birds and mammals, but 70-90per cent of most human populations are seropositive. The pathology of Parvoviruses is shaped by their dependence on cellular functions for replication - cell tropism is broad, but since cells must pass through S-phase, they tend to infect rapidly dividing tissues, most commonly:

- The foetus
- The intestinal epithelium
- The haematopoetic system

The only known human Parvovirus is referred to as B19. First discovered in 1974, it was only in 1981 that its association with aplastic crisis in children was first realized.

Type III: dsRNA

Type III viruses include:

- Enveloped phages (*Cystoviridae*),
- The animal- plant- and insect-infecting *Reoviridae,*
- The vertebrate- and invertebrate-infecting *Birnaviridae*
- The *Totiviridae,* in primitive Eukarya (fungi and protozoa, and perhaps in insects)
- *Partitiviridae,* which mainly infect fungi, and the
- Cryptoviruses (family *Partitiviridae*), which occur in plants, transmissible only via seed or pollen.

The viruses have:

- Single-component (*Totiviridae*)
- Two-component (*Birna-, Crypto-* and *Partitiviruses*)
- Three-component (*Cystoviridae*) and
- Multi-component (*Reoviridae,* 10-12 segments) genomes

All multiple component genomes are encapsidated in a single particle.

Virus genome sizes:

- *Partitiviridae*: 4-9 kb
- *Birnaviridae*: about 6 kb
- *Totiviridae*: 4-7 kb
- *Hypoviridae*: 10-13 kb
- *Cystoviridae*: 13 kb
- *Reoviridae*: 20-27 kb

All genomes apparently replicate by a conservative mechanism, wherein ds input RNA is transcribed by viral enzyme to mRNA, which both codes for protein, and acts as template for second strand synthesis.

Diarrhoea Viruses: Astroviruses; Caliciviruses; Reoviruses (Rotaviruses)

Diarrhoea is not as dramatic an illness as AIDS. However, the above groups of viruses are of great economic importance, causing millions of lost working days each year, as well as much discomfort. Diarrhoea continues to be a major cause of morbidity and mortality worldwide resulting in an estimated 1000 deaths among children each day, the highest incidence being in developing countries of the world. One well documented source of infection is the consumption of shellfish (polluted by sewage) - and therefore, they also have economic consequences for fishermen and the food industry. Such viruses often cause mini-epidemics in families, hospital wards, etc and are potentially very dangerous to seriously ill hospital patients. More importantly, these viruses contribute to the massive mortality caused by infantile diarrhoea in developing countries and are responsible for uncounted millions of deaths each year.

The human pathogens in these groups have been poorly studied since for the most part, they do not grow in culture. Only recently has molecular information on the nature of their genomes become available. The taxonomy of these viruses is still somewhat confused and they have previously been grouped on the basis of their appearance in the EM (unreliable). Recently, nucleotide sequences of members of each group have been determined, allowing genome organization to be used as a basis for future classification:

Rotaviruses

Infections cause endemic/epidemic gastroenteritis and infantile diarrhoea - second only to U.R.T. infections in infants <24 months - a major cause of death in the 3rd World:

Rotaviruses are the most common cause of severe diarrhoea worldwide. In developing countries, rotavirus infection may cause up to one million deaths each year, accounting for an estimated 20-25per cent of all deaths due to diarrhoea and 6per cent of all deaths among children less than five years old (Cook et al, Bull WHO, 1990; 68:171-7). Because of the magnitude of disease associated with rotavirus infections and because public health interventions to provide clean water and improved sanitation are unlikely to decrease the incidence of disease, vaccines are being developed as the first strategy for prevention. The first rotavirus vaccine, a live oral tetravalent vaccine derived from a rotavirus isolated from a rhesus monkey is expected be licensed in the United States and Europe in 1998.

Type IV: (+)ve-sense ssRNA

Class IV viruses include:

- Viruses infecting Eubacteria (*Leviviridae*)
- Viruses of insects (*Noda-, Tetraviridae*)
- Viruses of fungi (*Barnaviridae*)
- Viruses of plants (*Bromo-, Como-, Poty-, Sequi- and Tombusviridae,* and 19 unassigned genera)
- Viruses of vertebrates (*Astro-, Calici-, Corona-, Flavi-, Picorna- and Togaviridae,* and genus Arterivirus)

The viruses have:

- Single-component genomes with single ORFs (*Poty-, Picorna-, Sequiviridae*)
- Single components with multiple ORFs (*Toga-, Caliciviridae,* Tobamovirus)
- Two components with single ORFs (*Como-, Noda-, Tetri- and Potyviridae* genus Bymovirus)
- Two components with multiple ORFs (Tobra-, Furo-, Enamovirus)
- Three components with multiple ORFs (Hordeivirus, *Bromoviridae*)

Genome sizes range from:

- Less than 5 kb (*Levi-, Barna-, Tombusviridae*), to
- 7-12 kb (*Astro-, Como-, Picorna-, Calici-, Bromo-, Poty-, Sequi-, Toga-, Flaviviridae,* Tobraviruses)
- To 13 kb (Arteriviruses) - 15.5 kb (Closteroviruses)
- To 20 - 30 kb (*Coronaviridae*)

Genome expression strategies range from:

- Expressing a single polyprotein from each genome component and proteolytically processing it (*Picorna-, Poty-, Sequi-, Como-, Nodaviridae*)

- Expressing proteins from 5'-proximal ORFs and from subgenomic messengers (Tobamoviruses, *Bromoviridae*)
- A mixture of the two strategies (*Togaviridae,* Tymoviruses, *Caliciviridae*)

Astroviruses

Astroviruses were first described in 1975 as a result of EM studies of an outbreak of diarrhoea in a maternity unit and a survey into the etiology of diarrhoea in childhood in a Scottish city. Subsequently astroviruses were experimentally transmitted to adult volunteers and adapted to growth in foetal LLCMK2 cells. Restrictions on the availability and use of foetal tissue limited the cultivation of astroviruses and therefore studies were confined to a few laboratories that access to electron microscopes. Surveys conducted using EM showed that astrovirus infection occurs world-wide and accounts for 2-8per cent of cases of diarrhoea in infants. A recent pilot study investigating the causes of infectious intestinal disease in England based on surveillance in GP practices showed that astroviruses were the most frequent viral cause (Roderick et al, Epidemiol Infect 1995:114: 277-288). A major advance in the ability of laboratories to diagnose astrovirus came as the result of the finding that they could be propagated in a continuous line of colon carcinoma cells $CaCO_2$. The development of enzyme immunoassays (EIAs) for detecting viruses in stools showed that astrovirus is a significant cause of diarrhoea in developing countries. In Thailand, 8.6per cent of cases of diarrhoea were associated with astrovirus and a similar rate was found in Guatemala (7.3per cent). Application of EIA to study outbreaks in day care centres in the USA showed that many more cases were detected by EIA than by electron microscopy, the most widely used diagnostic technique.

The name of the family is derived from their distinctive appearance in electron microscopy - a smooth or slightly indented outer electron-dense shell with an inner, negatively staining 5 or 6 pointed star shaped core.

Clinically, these viruses cause similar symptoms to caliciviruses. Astroviruses have been isolated from birds, cats, dogs, pigs, sheep, cows and man. There are at least 7 human astrovirus serotypes. Like rotaviruses, astrovirus infections occur through the year with peaks in the winter months. Infections have been shown to occur mainly in childhood. A serological survey of 87 with children for antibody prevalence to astrovirus in England found that 67per cent of the children became seropositive by the age of 4 years and 87per cent by age 10. Other studies showed that most of the cases of infection were detected in children under five years of age with the majority of the children being under 1 year of age. Outbreaks of astrovirus infection involving children and elderly patients have been described and prolonged excretion documented in immunosuppressed, immunodeficient and AIDS patients. Significantly higher seroprevalence rates of astrovirus have been reported in adults exposed to contaminated water compared with a control group (Myint et al, Lancet 1995; 343:1640-1). Studies have shown that astrovirus type 1 is the prevalent strain in the UK accounting for 65per cent of the cases. However this situation may be changing with the emergence of more cases of astrovirus type 4. In contrast, type 2 was the most prevalent in Mexico (31per cent) and type 1 relatively rare (6per cent) and there are reports of outbreaks of foodborne astrovirus infection in Japan involving thousands of children and adults associated with astrovirus type 6.

Type V: (-)ve-sense ssRNA

Type V viruses include:

- Viruses infecting vertebrates only (*Arena-, Orthomyxo-, Paramyxoviridae*);
- Viruses infecting vertebrates and arthropods (*Bunya-, Rhabdoviridae*);
- Viruses infecting plants and arthropds (B*unya-, Rhabdoviridae*);
- Viruses infecting plants only (Tenuiviruses)

The viruses have:

- Single-component genomes with multiple ORFs (*Filo-, Paramyxo-, Rhabdoviridae*)

- Two-component ambisense genomes (Arenaviruses)
- Three-component, occasionally ambisense genomes (*Bunyaviridae,* Tenuivirus (4))
- Six to eight component genomes (*Orthomyxoviridae*)

The families *Paramyxo-, Rhabdo and Filoviridae* are grouped together in the Order Mononegavirales, due to similarities in virion structure, gene order and transcription.

Virus genome sizes:

- Single component, 11-19 kb: Mononegavirales
- 6 - 8 component, 10-14 kb: *Orthomyxoviridae*
- 3 component, 11-20 kb: *Bunyaviridae*

Class V viruses are probably a late evolutionary development, as they occur only among higher Eukarya (arthropods and vertebrates): viruses infecting plants probably do so as a result of close association of insects and host plants in recent evolutionary times.

Filoviruses

Name comes from the Latin: *filo* = 'threadlike'.

1967: Marburg/Frankfurt, Germany. Laboratory workers preparing primary cell cultures from African green monkeys resulted in an outbreak of a previously unrecognised disease. Highly infectious; 31 cases, 7 deaths (some probably survived due the first theraputic administration(?) of interferon).

This family has strong structural and genetic similarities to both Rhabdoviruses and the Paramyxoviruses.

Particles

Pleiomorphic, elongated, 80nm diameter × 130-14,000nm long - sometimes straight, but may be curved or hooked - 'U' or '6' shaped. Particles are rounded at one end with a distended swelling at the other. The core has a striated appearance similar to that of Rhabdoviruses.

Genome

The filovirus genome is s/s, unsegmented, (-)sense RNA, ~19kb. Size varies with length of particle - optimum infectivity for Marburg ~790nm, Ebola ~970nm. Encodes 7 proteins from monocistronic mRNA complementary to vRNA. Signals and expression strategy reminiscent of Rhabdoviruses and Paramyxoviruses, but there are differences, e.g. some filovirus genes overlap: The ribonucleoprotein complex consists of the nucleoprotein (NP), the structural proteins VP30, VP35 and the polymerase, L. The location of VP40 and VP24 have not been accurately determined, but are thought to be membrane associated. The L protein is the largest protein, and is the virion associated RNA dependent RNA polymerase. The glycoprotein (GP/SGP) is an integral membrane protein and is the only glycosylated virion protein, containing both N- and O-linked oligosaccharides.

Pathogenesis

The clinical manifestations of Ebola virus infection are severe. The incubation period varies between four and sixteen days. The initial symptoms are a severe frontal and temporal headache, generalised aches and pains, malaise, by the second day the victim will have a fever. Later symptoms include watery diarrhoea, abdominal pain, nausea, vomiting, a dry sore throat, and anorexia. By day seven of the symptoms, the patient will have a maculopapular (small slightly raised spots) rash. At the same time the person will develop thrombocytopenia and haemorrhagic manifestations, particularly in the gastrointestinal tract, and the lungs, but it can occur from any orifice, mucous membrane or skin site. By day twelve the skin starts to peel away from the rash. Ebola causes lesions in almost every organ, although the liver and spleen are the most noticeably affected. Both are darkened and enlarged with signs of necrosis. The cause of death is normally shock, associated with fluid and blood loss into the tissues.

The haemorrhagic and connective tissue complications of the disease are not really understood, but may be related to the fact that the VP40 protein is antigenically related to human cell matrix proteins (abdominal aortic aneurism protein and MFAP-4), leading to autoimmune attack.

Why does the immune system not clear the infection? This may be associated with the two forms of the virus glycoprotein.

The glycoprotein gene has a translation stop codon in the middle of it, preventing the synthesis of the full length protein. Approximately twenty percent of the mRNA isolated from infected cells had been edited to contain an extra adenosine in a stretch of seven adenosine residues at positions 1019-1026. This causes a frame shift, allowing the synthesis of the full length protein. The larger protein (130Kd - GP) is membrane associated protein, and the truncated version (approximately 60 Kd - SGP) is secreted.

A possible role for SGP is to protect the virus from the immune system as a decoy antigen. However, SGP binds to neutrophils and interferes with their function. Moreover, GP also appears to be immunosuppressive, further interfering with the response to infection.

Relatively little work has been performed on these viruses because of the difficulties of working with them, but it is now known that:

- They are possibly simian viruses, probably widespread throughout Africa, although no animal reservoir has yet been identified.
- Serological surveys in Africa suggest that asymptomatic human infections occur sporadically (i.e. subjects still alive!).

Paramyxoviruses

The family is divided into 3 genera:

Paramyxovirus: Parainfluenzavirus 1-4; Mumps

Pneumovirus: Respiratory Syncytial Virus (RSV)

Morbillivirus: Measles; Canine Distemper Virus.

Parainfluenzaviruses and RSV produce acute respiratory diseases (c.f. influenza), Morbilliviruses and Mumps systemic disease - diversity! They also differ from Orthomyxoviruses genetically - non-segmented genome with little genetic variation (c.f. influenza).

Morphology

Glycoproteins: Do not form such prominent spikes as on influenza virus:

HN: Haemagglutinin + neuraminidase activities;

Measles: Referred to as H protein - no neuraminidase activity;

RSV: G protein - neither activity.

F: Consists of 2 disulphide-linked subunits (F1 + F2) - responsible for cell fusion + haemolytic function.

Other proteins:

The M (matrix) protein lines the inner surface of the envelope.

NP: Nucleoprotein.

L and P: Polymerase activity.

Genome

Non-segmented (-)sense RNA, 17-20kb. The linear arrangement of genes (6) are separated by repeated sequences, a polyadenylation signal at the end of the gene, the intergenic sequence GAA followed by a translation start signal at the beginning of the next gene.

Replication

Very similar for all viruses in this group. Unlike influenza, all the action occurs in the cytoplasm. However, the overall strategy very similar to influenza, although unlike influenza, Paramyxovirus replication is resistant to actinomycin D.

A large excess of nucleocapsids are produced in infected cells, which form characteristic cytoplasmic inclusion bodies. Syncytium formation is quite common (F glycoprotein).

Pathogenesis

Parainfluenzaviruses 1-4: Cause acute respiratory infections of man ranging from relatively mild influenza-like illness to bronchitis, croup (narrowing of airways which can result in respiratory distress) and pneumonia; common infection of children. Transmitted by aerosols, virus is usually limited to U.R.T. (no viraemia). Infections of L.R.T. (e.g. in very young children) lead to more serious symptoms. Little serological variation, therefore rare infection in adults.

Mumps

Recognised by the ancient Greeks, virus first isolated in 1934. Haemagglutination is a valuable assay technique for this virus. Humans are believed to be the only natural reservoir for the virus (possibly primates). Transmission via saliva and respiratory secretions; less infectious than measles/chickenpox - more adult cases. Typically causes painful swelling of parotid glands 16-18 days after infection. This is preceded by primary replication of the virus in epithelial cells of the U.R.T. and local lymph nodes, followed by viraemia. In children, mumps is usually self-limited, but in adults (post-puberty) a proportion of cases have more serous sequalae: orchitis (20-30per cent of males - rarely resulting in sterility); meningitis, encephalitis, pancreatitis, myocarditis, nephritis - <1per cent adult cases.

Treatment: none (passive immunization has been used).

Prevention: one invariant serotype therefore vaccines are viable - both formalin-inactivated and live attenuated exist, the latter now being widely used.

Measles

One of the most infectious diseases known! >10^6 deaths p.a. in children in the third world - now part of the W.H.O expanded programme of immunization. Childhood infection almost universal, protection resulting from this is probably lifelong. Both man and wild monkeys are commonly infected, but the virus can also infect rodents (in wild?). In culture, produces characteristic intranuclear inclusion bodies and syncytial giant cells. Transmission and initial stages of disease similar to mumps, but this virus can also infect via the eye and multiply in the conjunctivae. Viraemia following primary local multiplication results in widespread distribution to many organs.

Symptoms: After a 10-12 day incubation period, dry cough, sore throat, conjunctivitis (virus may be excreted during this phase!!!), followed a few days later by the characteristic red, maculopapular rash and Koplik's spots - raised red spots with white centres in the mouth. Towards the end of the disease, there is extensive, generalized virus infection in lymphoid tissues and skin. Complications include bronchopneumonia and otitis media (with or without secondary bacterial infections) (relatively common), and encephalitis (~1:2000 cases). Subacute schlerosing pan encephalitis (SSPE) results from a rare (~1: $3x10^5$ cases of measles), chronic infection in which the virus multiplies in the brain with the expression of a limited repertoire of virus genes, resulting in neurodegenerative disease.

Treatment: None

Prevention: Both live and killed vaccines exist. Vaccination with the live attenuated vaccine has been practiced in the US since the 1960's with a dramatic decline in the incidence of the disease (210), but has only been used more recently in the UK. Trivalent live attenuated vaccine (MMR) usually given - all of these viruses best avoided during pregnancy!

RSV

First isolated in 1956 and subsequently recognised as a major cause of L.R.T. disease in infants and young children. Infects man, monkeys and some rodents with disease production, but inapparent infections (resulting in spread of virus) may occur in many mammals. In culture, causes characteristic syncytial masses - hence the name. Highly infectious, transmission by respiratory secretions. Primary multiplication occurs in epithelial cells of U.R.T. producing a mild illness. In ~50per cent children less than 8 months old, virus subsequently spreads into the L.R.T. causing bronchitis, pneumonia and croup. Has been suggested as a possible factor in cot death. Prevention: Currently no vaccine! Also, infection does not result in lasting protection (c.f. mumps, measles) therefore repeated infections ('colds') occur throughout life - usually without serious consequences in adults.

Rhabdoviruses

Genetically, these viruses have non-segmented (–)sense NA genome reminiscent of Paramyxoviruses. There are >200

Rhabdoviruses known (probably still an underestimate of the total), which infect man (Rabies - the only member of the group to do so 'naturally'), other mammals, fish, insects (some replicate in arthropods and were previously classified as Arboviruses) and plants - versatile! The family is split into 5 genera:

Lyssavirus (Greek. 'frenzy')	:	Rabies virus; other mammalian viruses; some insect viruses
Vesiculovirus	:	Vesicular Stomatitis Virus (VSV);
mostly insect viruses		
Ephemerovirus	:	Bovine ephemeral fever virus
(vertebrates)		
Cytorhabdovirus	:	Lettuce necrotic yellows virus (plants)
Nucleorhabdovirus	:	Potato yellow dwarf virus (plants)

Rabies is an 'ancient' disease, first shown to be of infectious origin in 1808, shown to be of viral etiology by Pasteur in the 1880's (when Pasteur and Koch were developing the germ theory of disease - prior to the firm modern definition of 'viruses' by Beijerinick (1898)). Over a decade, Pasteur carried out the serial passage of Rabies virus in rabbits, and eventually succeeded in isolating an attenuated preparation which was used to treat patients bitten by mad dogs (not without some risks).

Morphology

Particles ca. 180 × 70nm with unique bullet-shaped appearance (all rather similar). Enveloped with prominent spikes on surface (G protein - haemagglutinates RBCs), but not very variable in appearance. The envelope is lined by the matrix protein and contains the nucleocapsid (RNA + N protein) wound helically inside the core. Two non-structural proteins, L and NS, are associated with the nucleocapsid and act in concert as the viral polymerase. As with most enveloped viruses, the particle is relatively labile.

Genome

~11kb. ~50nt 'leader' region at the 3' end and ~60 non-translated region at the 5' end of the (-)sense vRNA. Similar to paramyxoviruses, there is a conserved polyadenylation signal at the end of each gene and a short intergenic region between each of the 5 genes. Most of our knowledge of Rhabdoviruses comes from Vesicular Stomatitis Virus (VSV) - very similar to Rabies, but the entire rabies genome has now been cloned, sequenced and expressed.

Replication

Transmission varies depending on virus/host, but most are transmitted by direct contact - e.g. rabies - animal bites or insect vector. There is a long incubation period in vivo, but this is not reflected in the kinetics of virus replication in culture. The G protein spikes bind to receptors on the surface of host cells and the viruses enters the cell by endocytosis and fusion with the membrane of the vesicle (as Paramyxoviruses), mediated by the G protein.

Receptor molecules for Rhabdoviruses are not known, but are believed to be phospholipids rather than specific proteins. Replication occurs in the cytoplasm - both the L and NS proteins are necessary for transcription - neither function alone. Five monocistronic mRNAs are produced, capped at the 5' end and polyadenylated at the 3' end and each containing the leader sequence from the 3' end of the vRNA at the 5' end of the message. These mRNAs are made by sequential transcription of the ORFs in the virus genome and it has been shown that the intergenic sequence is responsible for termination and re-initiation of transcription by the polymerase between each gene, thus producing separate transcripts.

Progeny vRNA is made from a (+)sense intermediate, but this step is not well understood. The genome is replicated by the L + P polymerase complex (as in transcription), but additional host cell factors (not known) are also required. It is characteristic of Rhabdoviruses that these events all occur in a portion of the cytoplasm which acts as a virus 'factory' and appears as a characteristic cytoplasmic inclusion body - e.g. Rabies - perinuclear Negri bodies.

Virions are assembled around the tightly coiled nucleoprotein core, and bud both from cytoplasmic membranes and the outer membrane of the cell (acquiring the M + G proteins as they do so).

Pathogenesis: Rabies

Entry occurs by wound or abrasion of skin directly into bloodstream (animal bite - e.g. vampire bats in S. America). 1o replication occurs locally in muscle and connective tissue (no symptoms), but virus eventually infects peripheral nerves, then travels (passively?) along neuronal axons to CNS, where it produces severe and fatal encephalitis. Few cases escape these severe consequences. Viraemia and haematogenous spread of virus to CNS has not been shown. Incubation period varies from 3-8 weeks to 1 year depending on size and site of inoculation (e.g. head/face/neck vs. hands or feet).

Epidemiology

Man is a dead-end infection for the virus, normal hosts are foxes, dogs, cats, bats, skunks, etc. Cattle rabies is a serious economic disease in countries where Rabies is endemic. Pattern is endemic rather than epidemic.

Treatment

No effective drug treatment, but passive immunization is of value. This is one of the cases where therapeutic (post-exposure) vaccination is important - the aim is not to prevent infection but to moderate the severity of the disease. 3 types of vaccine exist:

- Inactivated: as pioneered by Pasteur in the 1880's - used until the 1950's (material from infected animals containing up to 5per cent nervous tissue - provoked severe immune reactions in recipients, but saved lives). This type of vaccine was then produced by growing the virus in embryonated duck eggs - better, but still produced severe encephalitis in some recipients. Inactivated vaccine is now produced by growing the virus in diploid human fibroblasts -

effective and safe but expensive, therefore duck egg-produced material still used in third world.

- Live attenuated: has been used to vaccinate domestic animals and livestock. Effective, but not considered sufficiently safe for human use.
- Recombinant: recently developed recombinant vaccinia virus / G protein vaccine is being used to eradicate rabies in foxes in Europe

VSV

Normally causes an epidemic, self-limited disease of cattle which is of some economic importance in countries where it is endemic, e.g. N.America. Similar in appearance to foot and mouth disease! Can infect man, causing a relatively mild febrile illness.

Retroid Genome Types

A) Type VI: Diploid ssRNA Genomes Replicating via Longer-Than-Genome-Length dsDNA Intermediates

Class VI viruses include:

Only viruses of vertebrates in the family *Retroviridae*

These are (by genus):

- Mammalian type B retroviruses (eg: mouse mammary tumour virus, MMTV)
- Mammalian type C retroviruses (Moloney murine leukaemia virus, MuLV)
- Avian type C viruses (avian leukaemia virus, ALV)
- Type D viruses (Mason-Pfizer monkey virus, MPMV)
- BLV-HTLV viruses (human T-lymphotropic virus, HTLV)
- Lentivirus (HIV, SIV)
- Spumavirus (human spumavirus)

The virus genomes are:

- Diploid ssRNA (held together by protein)
- Messenger sense

- Capped and poly-A tailed
- With 4 basic coding regions:

Retroviridae replicate via a dsDNA, longer-than-genome-length intermediate (provirus), which is integrated covalently into the host cell chromosomal DNA.

Conversion of RNA to DNA, and integration into host DNA, is done by the viral-coded RNA-dependent DNA polymerase, which also has RNAse H activity, and DNA-dependent DNA pol activity, as well as encoding the integrase function. Transcription of viral mRNA - which is spliced to allow expression of 3'-proximal ORFs - is by host RNA pol II.

B) Type VII: dsDNA Genomes Replicating via Longer-Than-Genome-Length ssRNA Intermediates

Class VII viruses include:

Viruses with encapsidated circular dsDNA genomes (usually gapped or nicked), which replicate via longer-than-genome-length messenger-sense ssRNA intermediates. Such as:

- Viruses infecting vertebrates (*Hepadnaviridae*)
- Isometric viruses infecting plants (Caulimoviruses)
- Bacilliform viruses infecting plants ("Badnaviruses")
- And now an infectious *gypsy*-type retrotransposon of *Petunia* (petunia vein-clearing virus)

The viruses have:

- Partially double-stranded non-covalently closed circular DNA, 3.2 kb (*Hepadnaviridae*)
- Circular dsDNA with single-strand break(s) in both strands, about 8 kb (Caulimoviruses and Badnaviruses and petunia virus)

All the viruses are probably derived from the same origin as each other, which they share with Retroviridae: for this reason Class VII viruses have been proposed as "Pararetroviridae".

After the RNA retrovirus enters a host cell, its genomic RNA will be transcribed into a double stranded DNA and then integrated into the host DNA. The RNA to DNA transcription is called reverse transcription.

The entire process is catalyzed by reverse transcriptase which has both DNA polymerase and RNase H activities.

1. A retrovirus-specific cellular tRNA hybridizes with a complementary region called the primer-binding site (PBS).
2. A DNA segment is extended from tRNA based on the sequence of the retroviral genomic RNA.
3. The viral R and U5 sequences are removed by RNase H.
4. First jump: DNA hybridizes with the remaining R sequence at the 3' end.
5. A DNA strand is extended from the 3' end.
6. Most viral RNA is removed by RNase H.
7. A second DNA strand is extended from the viral RNA.
8. Both tRNA and the remaining viral RNA are removed by RNase H.
9. Second jump: The PBS region of the second strand hybridizes with the PBS region of the first strand.
10. Extension on both DNA strands. LTR stands for "long terminal repeat".

GENERATION OF RETROVIRAL GENOMIC RNA

Replication of RNA retroviruses requires generation of retroviral genomic RNA from the integrated DNA. This involves the regular DNA to RNA transcription. The retroviral DNA contains two long terminal repeats (LTR), but only the upstream LTR directs transcription. Another LTR directs the cleavage of the primary transcript to add a poly-A tail.

It contains the recognition sites of several transcription factors in human cells. Transcription starts from the left end of the R region.

The primary transcript includes the right LTR which can direct the cleavage at the last nucleotide of the R region to add the poly-A tail.

Chapter 5

Post-Transcriptional Processes

RNA PROCESSING

RNA processing is to generate a mature mRNA (for protein genes) or a functional tRNA or rRNA from the primary transcript. In this section, we discuss first the processing of pre-mRNA and then processing of pre-rRNA and pre-tRNA.

Processing of pre-mRNA involves the following steps:

- Capping: Add 7-methylguanylate (m^7G) to the 5' end.
- Polyadenylation: Add a poly-A tail to the 3' end.
- Splicing: Remove introns and join exons.

In some cases, RNA editing is also involved.

5'-Capping

Capping occurs shortly after transcription begins. The chemical structure of the "cap" is shown in the following figure, where m^7G is linked to the first nucleotide by a special 5'-5' triphosphate linkage. In most organisms, the first nucleotide is methylated at the 2'-hydroxyl of the ribose. In vertebrates, the second nucleotide is also methylated.

3'-Polyadenylation

A stretch of adenylate residues are added to the 3' end. The poly-A tail contains ~ 250 A residues in mammals, and ~ 100 in yeasts.

Processing of pre-rRNA and pre-tRNA

The newly transcribed pre-rRNA is a cluster of three rRNAs: 18S, 5.8S and 28S in mammals. They must be separated to become functional. Pre-rRNA is synthesized in the nucleolus. The U3 snRNA, other U-rich snRNAs, and their associated proteins in the nucleolus are involved in the cleavage of the pre-rRNA.

5S rRNA is synthesized in the nucleoplasm. It does not require any processing. After 5S rRNA is synthesized, it will enter the nulceolus to combine with 28S and 5.8S rRNAs, forming the large subunit of the ribosome.

Pre-tRNA requires extensive processing to become a functional tRNA. Four types of modifications are involved:

1. Removing an extra segment (~ 16 nucleotides) at the 5' end by RNase P.
2. Removing an intron (~ 14 nucleotides) in the anticodon loop by splicing.
3. Replacing two U residues at the 3'end by CCA, which is found in all mature tRNAs.
4. Modifying some residues to characteristic bases, e.g., inosine, dihydrouridine and pseudouridine.

RNA SPLICING

RNA splicing is a process that removes introns and joins exons in a primary transcript. An intron usually contains a clear signal for splicing (e.g., the beta globin gene). In some cases (e.g., the sex lethal gene of fruit fly), a splicing signal may be masked by a regulatory protein, resulting in alternative splicing. In rare cases (e.g., HIV genes), a pre-mRNA may contain several ambiguous splicing signals, resulting in a few alternatively spliced mRNAs.

Splicing signal

Most introns start from the sequence GU and end with the sequence AG (in the 5' to 3' direction). They are referred to as the splice donor and splice acceptor site, respectively.

However, the sequences at the two sites are not sufficient to signal the presence of an intron. Another important sequence is called the branch site located 20 - 50 bases upstream of the acceptor site. The consensus sequence of the branch site is "CU(A/G)A(C/U)", where A is conserved in all genes.

In over 60per cent of cases, the exon sequence is (A/C)AG at the donor site, and G at the acceptor site.

Exception for the GU-AG rule is discussed in the following review article:

AT-AC Pre-mRNA Splicing Mechanisms and Conservation of Minor Introns in Voltage-Gated Ion Channel Genes - Molecular and Cellular Biology, 1999.

Splicing Mechanism

The detailed splicing mechanism is quite complex. In short, it involves five snRNAs and their associated proteins. These ribonucleoproteins form a large (60S) complex, called spliceosome. Then, after a two-step enzymatic reaction, the intron is removed and two neighboring exons are joined together. The branch point A residue plays a critical role in the enzymatic reaction.

RNA EDITING

RNA editing is a molecular process in which the base of a RNA molecule is altered by specific enzymes. In the following example, Cytosine (C) is changed to Uracil (U).

NUCLEAR TRANSPORT

After RNA molecules (mRNA, tRNA and rRNA) are produced in the nucleus, they must be exported to the cytoplasm for protein synthesis. On the other hand, many proteins operating in the nucleus must be imported from the cytoplasm. The traffic through the nuclear envelope is mediated by a protein family which can be divided into exportins and importins. Binding of a molecule (a "cargo") to exportins facilitates its export to the cytoplasm. Importins facilitate import into the nucleus.

The function of exportins and importins is regulated by a G protein called "Ran". There are two types of G proteins: heterotrimeric G proteins and monomeric G proteins (or small G proteins). The latter includes Ras, Ran, Rho, Rab, etc. Like other G proteins, Ran can switch between GTP-bound and GDP-bound states. Transition from the GTP-bound to the GDP-bound state is catalyzed by a GTPase-activating protein (GAP) which induces hydrolysis of the bound GTP. The reverse transition is catalyzed by guanine nucleotide exchange factor (GEF) which induces exchange between the bound GDP and the cellular GTP.

The GEF of Ran (denoted by RanGEF) is located predominantly in the nucleus while RanGAP is located almost exclusively in the cytoplasm. Therefore, in the nucleus Ran will be mainly in the GTP-bound state due to the action of RanGEF while cytoplasmic Ran will be mainly loaded with GDP. This asymmetric distribution has led to the following model for the function of exportins and importins.

It is thought that binding between an exportin or importin and its cargo depends on their interaction with Ran: RanGTP enhances binding between an exportin and its cargo but stimulates release of importin's cargo; RanGDT has the opposite effect, namely, it stimulates the release of exportin's cargo, but enhances the binding between an importin and its cargo. Therefore, the exportin and its cargo may move together with RanGTP inside the nucleus, but the cargo will be released as soon as the complex moves into the cytoplasm (through nuclear pores), since RanGTP will be converted to RanGDP in the cytoplasm. By contrast, the importin and its cargo may move together with RanGDP in the cytoplasm, but the cargo will be released in the nucleus since RanGDP will be converted to RanGTP in the nucleus.

THE ROLE OF HIV REV PROTEIN

The major role of rev is to regulate the expression of HIV proteins by controlling the export rate of mRNAs. The HIV mRNAs are produced from the primary transcript by three different splicings: unspliced, singly spliced and doubly spliced. Although unspliced and singly sliced mRNAs are made before doubly spliced mRNAs, the protein products of the doubly spliced mRNAs are the first synthesized in the cytoplasm because they are smaller and exported faster. As a matter of fact, expression of unspliced and singly sliced mRNAs would be negligible without the rev protein which is a product of the doubly spliced mRNA.

a. The rev protein is a product of the doubly spliced mRNA. Without rev, export of unspliced and singly spliced mRNAs (I and II) is very slow.
b. The rev protein can bind to the rev-response-element (RRE) of mRNA I and II, accelerating their export.
c. Sequences of NLS and NES in the rev proten.

A rev protein consists of 116 amino acids, including a nuclear localization sequence (NLS) and a nuclear export sequence (NES). The 3D structure of rev is not known yet. However, the hydrophilic NLS is likely to be more exposed than the hydrophobic NES. Therefore, when rev is synthesized in the cytoplasm, the importin α/β may bind to its NLS, carrying it into the nucleus. Even if NLS and NES are equally exposed, the action of Ran still favors the binding of the cytoplasmic rev to importin, rather than exportin.

When rev is in the nucleus, it can bind to mRNA's RRE. The RNA binding domain of rev is in the same region as NLS. Thus, upon binding to mRNA, the NLS of rev is masked, exposing only NES. This facilitates the export of the rev/mRNA complex by the exportin Crm1.

NFAT, CALCINEURIN AND IMMUNOSUPPRESSION

Organ transplantation often elicits immune responses to reject the foreign object. The most widely used drugs to deal with such graft reaction are FK506 and cyclosporin A (CsA). Their direct target is a phosphatase known as calcineurin. CsA combines with a cellular protein called cyclophilin to inhibit calcineurin; FK506 combines with FKBP (FK506 binding protein) to inhibit the phosphatase.

How could inhibition of a phosphatase suppress the immune response? The answer lies in the transport of NFAT (nuclear factor of activated T cells) which is a transcription factor regulating IL-2.

NFAT contains both nuclear localization sequence (NLS) and nuclear export sequence (NES), but one of them is buried in the protein interior, inaccessible to importin and exportin. Whether NLS or NES is masked depends on the phosphorylation state of specific serine residues in the regulatory domain. Phosphorylation of these serine residues exposes NES whereas dephosphorylation exposes NLS.

In resting cells, the NLS of the cytoplasmic NFAT is masked due to phosphorylation on these serine residues. In stimulated cells, an increase of intracellular calcium ions activates calcineurin, which then dephosphorylates the masking residues. Consequently, NLS is exposed and NFAT can be carried into the nucleus by the importin a/b. Inside the nucleus, NFAT may be re-phosphorylated by protein kinase, exposing its NES and be exported by the exportin Crm1.

PROTEIN SYNTHESIS OVERVIEW

Protein synthesis is carried out on ribosomes based on the sequence of mRNA. It always starts from methionine, encoded by the codon AUG. However, a polycistronic mRNA should contain multiple initiating codons. On the other hand, a peptide may also contain several non-initiating methionine residues, also encoded by AUG. How could the system distinguish them? The answer lies in the initiation signals.

The synthesized peptide sequence is a translation of mRNA sequence according to the genetic code. It starts from the initiation codon, and then follows the mRNA sequence in a strictly "three nucleotides for one amino acid" manner. Therefore, a minor change in the mRNA sequence could produce a very different peptide. For instance, if a codon which codes for an amino acid is changed to a "stop" codon, the subsequent sequence will not be translated. Another example is the "frameshift".

The translation is carried out by tRNA (more info). The entire procedure of protein synthesis includes initiation, elongation and termination (more info).

INITIATION SIGNALS OF PROTEIN SYNTHESIS

Synthesis of a peptide always starts from methionine. However, the first amino acid of a functional protein is not always methionine. After a peptide is synthesized, it may undergo a variety of modifications, including the cleavage of its N-terminus. Prokaryotes and eukaryotes use different mechanisms to recognize the initiating codon.

Prokaryotes

Many prokaryotic mRNAs are polycistronic, namely, an mRNA encodes more than one peptide chain. The polycistronic mRNA should contain multiple initiating codons. However, a peptide may also contain several non-initiating methionine residues. In the standard genetic code, the codon for both initiating and non-initiating methionine is AUG. To distinguish them, prokaryotes use a specific sequence located about 5-10 bases upstream of the initiation AUG. The specific sequence, UAAGGAGG, is known as the Shine-Dalgamo sequence (after its discoverers).

The 16S rRNA of the ribosome contains a sequence which can pair with the Shine-Dalgamo sequence:

5'—UAAGGAGG(5-10 bases)AUG mRNA

3'—AUUCCUCC........ 16S rRNA

Their association will recruit other parts of the ribosome to the initiation site for protein synthesis.

Eukaryotes

The mechanism used by eukaryotes to recognize the initiating AUG is not entirely clear. Evidence suggests that the eukaryotic ribosome may simply scan from the 5' cap and identify the first AUG as the initiation site. This mechanism is reasonable because nearly all eukaryotic mRNAs are monocistronic (encodes a single peptide). However, some viral mRNAs are polycistronic or lack 5' cap. It is not fully understood how the host eukaryotic cell recognizes the initiation sites of these viral mRNAs. Marilyn Kozak found that the following sequence may increase the effectiveness as an initiation site:

5'—ACCAUGG – mRNA

FRAMESHIFT

In most cases, frameshift involves the insertion or deletion of a single nucleotide in mRNA. Theoretically, it could involve more than one nucleotide, as long as the number is not a multiple of 3. When a nucleotide is added to or deleted from the mRNA, the subsequent sequence will produce an entirely different peptide.

mRNA(a) and mRNA(b) differ by only one nucleotide: mRNA(b) has an additional nucleotide "G" at the third position in this figure. Note that the translated amino acids are entirely different after the insertion point.

GENE OVERLAPPING

From the phenomenon of frameshift, we see that the same nucleotide sequence could encode different amino acid sequences. This allows two genes with entirely different protein products to share the same nucleotide sequence. The gene overlapping is often found in a compact genome such as HBV and HIV. In human mitochondrial DNA, there are two overlapping genes: ATPase subunits 6 and 8.

TRANSLATION BY TRNA

Translation is carried out by tRNA through the relationship between its anticodon and the associated amino acid. When a tRNA is brought to the ribosome by the pairing between its anticodon and the mRNA's codon, the amino acid attached at its 3' end will be added to the growing peptide. In bacteria, there are 30-40 tRNAs with different anticodons. In animal and plant cells, about 50 different tRNAs are found. However, there are 61 codons coded for amino acids. Suppose each codon can pair with only a unique anticodon, then 61 tRNAs would be needed.

Wobble Pairing

The reason why less than 61 tRNAs are required is because of the "wobble pairing" between anticodon and codon. As shown in the following figure, base pairing does not obey the standard Watson-Crick pairing at the wobble position. One base can pair with several other bases.

Aminoacyl-tRNA

The tRNA together with the amino acid attached to its 3' end is called an aminoacyl-tRNA. The attached amino acid is encoded by the codon which matches the tRNA's anticodon. Therefore, only a specific amino acid may be attached to the 3' end of a given tRNA. The process is catalyzed by a class of enzymes called aminoacyl-tRNA synthetases, which recognize the anticodon and its compatible amino acid. A cell has 20 different aminoacyl-tRNA synthetases, each can add only one of 20 amino acids to a compatible tRNA.

Let us use arginine as an example to explain the notation of aminoacyl-tRNA. "Arg-tRNA" denotes the tRNA with arginine attached while "$tRNA^{Arg}$" specifies the tRNA which contains the anticodon for the codon of arginine. "Arg-$tRNA^{Arg}$" represents the arginine-specific tRNA with arginine attached.

PROCEDURE OF PROTEIN SYNTHESIS

Initiation

Peptide synthesis always starts from methionine (Met). Therefore, the initial aminoacyl-tRNA is Met-$tRNA_i^{Met}$, where the subscript "i" specifies "initiation". In bacteria, the methionine of the initial aminoacyl-tRNA has been modified by the addition of a formyl group (HCO) to its amino group. The modified methionine is called formylmethionine (fMet), which is unique for bacteria. Thus, fMet is an obvious foreign substance in eukaryotes. It can elicit a strong immune response. In humans, the immune response elicited by the peptide "fMet-Leu-Phe" is about a thousand times greater than "Met-Leu-Phe".

Elongation

A ribosome contains two major tRNA-binding sites: A site and P site. After the large subunit joins the initiation complex, the initial Met-$tRNA_i^{Met}$ enters the P site and the newly arrived aminoacyl-tRNA is always placed at the A site ("A" for "aminoacyl"). Then, methionine is transferred to the new aminoacyl-tRNA, forming a "peptidyl-tRNA" where a peptide is attached to the tRNA. Subsequently, the empty tRNA at the P site is ejected from the ribosome and the peptidyl-tRNA jumps to the P site ("P." for "peptidyl"). During this translocation step, the ribosome also moves one codon down the mRNA chain. Similar steps are repeated in the next cycles of elongation.

Termination

Protein synthesis will terminate when the ribosome arrives at one of three stop codons. The termination process is assisted by special proteins called termination factors which recognize the stop codons. Their association stimulates the release of the peptidyl-tRNA from the ribosome. Subsequently, the released peptidyl-tRNA divides into tRNA and a newly synthesized peptide chain. The ribosome also divides into the large and small subunits, ready for synthesizing another peptide.

1. In the absence of mRNA, the large and small subunits of a ribosome are separated. At the beginning of peptide synthesis, initiation factors (IF) first assist the assembly of the small subunit, mRNA and the initial aminoacyl-tRNA. Then, the large subunit is recruited to join the complex.
2. In the elongation process, one cycle involves the following steps:
   i. New entry. A new aminoacyl-tRNA with a correct anticodon is brought to the A site. This step is catalyzed by elongation factors Tu and Ts in prokaryotes, and by elongation factors EF_1 and EF_{1b} in eukaryotes..

ii. Peptide synthesis. The peptide attached to the peptidyl-tRNA at the P site is transferred to the new aminoacyl-tRNA at the A site, generating a peptidyl-tRNA with a longer peptide. This step is catalyzed by peptidyl transferase.

iii. Translocation. The empty tRNA at the P site is ejected from the ribosome and the peptidyl-tRNA generated at the A site takes over the vacant P site. In the mean time, the ribosome moves one codon down the mRNA chain. The A to P switch is catalyzed by the elongation factor G in bacteria, and by the elongation factor EF_2 in eukaryotes.

SORTING OF PROTEINS

As discussed in the previous section, proteins are synthesized on ribosomes which are located mainly in the cytosol. Only a small number of ribosomes are located in mitochondria and chloroplasts. Proteins synthesized on these ribosomes can be directly incorporated into the compartments within these organelles. However, most mitochondrial and chloroplast proteins are encoded by nuclear DNA and synthesized on cytosolic ribosomes. These and all other proteins synthesized in the cytosol must be transported to appropriate locations in the cell. This is made possible by the specific signal sequence in the newly synthesized peptide.

1. If the N terminus of the new peptide contains a stretch of hydrophobic residues, it is sent to the rough endoplasmic reticulum (ER) for further sorting. Otherwise, it goes to non-ER pathways.

2. The new peptide is retained in the rough ER if its C-terminus contains the sequence "Lys-Asp-Glu-Leu" (KDEL in one-letter code). Otherwise, it will move to the Golgi apparatus.

3. Proteins containing a specific transmembrane α helix will be localized in the Golgi apparatus.

4. After glycosylation at the Golgi apparatus, the modified protein containing a mannose 6-phosphate (M6P) will be delivered to the lysosome.
5. Proteins that can aggregate with chromogranin B (secretogranin I) or secretogranin II will be sorted into regulated secretory vesicles where proteins are released upon specific stimulation. Otherwise, they are sorted into another type of vesicles which continuously move to the plasma membrane or outside the cell.

TRANSPORT INTO ROUGH ER

Insertion

a. Before synthesis.

b. After about 70 amino acids have been synthesized, the signal sequence emerges from the ribosome. A signal recognition particle (SRP) directs the new peptide (together with the ribosome) toward the ER membrane surface, because it can bind both the new peptide and a receptor on the membrane.

c. The signal sequence inserts itself into the ER membrane with the help of a signal sequence binding protein. Several proteins are recruited to form a channel for the growing peptide to pass through the membrane.

d. SRP dissociates from the peptide-ribosome complex. The signal sequence is cleaved and quickly degraded.

Folding

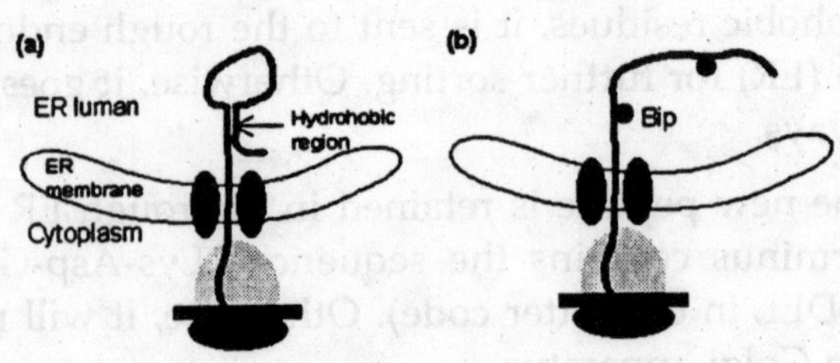

Fig. Prevention of Protein Misfolding by the Protein BiP (a Chaperone).

a. Before a peptide is completely synthesized, it may fold prematurely since hydrophobic regions tend to aggregate.

b. BiP can bind to unfolded, hydrophobic segments of peptides, preventing premature aggregation.

TRANSPORT INTO AND ACROSS GOLGI

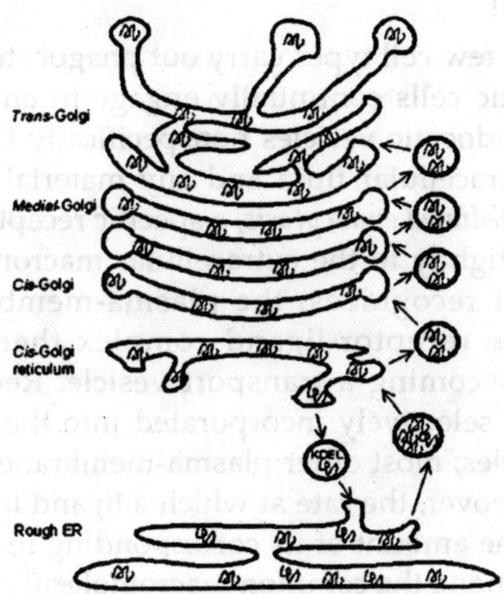

Fig. The Golgi Apparatus.

Transport of proteins into and across the Golgi apparatus is mediated by small vesicles.

TRANSPORT TO LYSOSOME

Glycosylated proteins in the Golgi apparatus will be delivered to lysosome if one or more mannose residues are phosphorylated on carbon atom 6. The *trans*-Golgi reticulum contains mannose 6-phosphate receptor which can bind to the phosphorylated mannose on lysosomal enzymes, directing the incorporation of the enzymes into transport vesicles.

ENDOCYTOSIS AND EXOCYTOSIS

In previous sections, we've followed the main pathways whereby secretory and membrane proteins are synthesized within cells and then targeted to the cell surface or other destination. However, cells also can internalize materials from their surroundings by phagocytosis, an actin-mediated process in which cells envelop bacteria and other large particles and then internalize them, and endocytosis, a process in which a small region of the plasma membrane invaginates to form a new intracellular membrane-limited vesicle about 0.05 to 0.1 μm in diameter.

Relatively few cell types carry out phagocytosis, whereas most eukaryotic cells continually engage in endocytosis. In pinocytosis, endocytic vesicles nonspecifically take up small droplets of extracellular fluid and any material dissolved in it. In *receptor-mediated endocytosis,* a specific receptor on the cell surface binds tightly to the extracellular macromolecule (the ligand) that it recognizes; the plasma-membrane region containing the receptor-ligand complex then undergoes endocytosis, becoming a transport vesicle. Receptorligand complexes are selectively incorporated into the intracellular transport vesicles; most other plasma-membrane proteins are excluded. Moreover, the rate at which a ligand is internalized is limited by the amount of its corresponding receptor on the cell surface. Among the common macromolecular ligands that vertebrate cells internalize by receptor-mediated endocytosis are cholesterol-containing particles called *low-density lipoprotein* (LDL); *transferrin,* an iron-binding protein; insulin and most other protein hormones; and glycoproteins whose oligosaccharide side chains contain terminal glucose, mannose, or galactose residues rather than the normal sialic acid.

Small invaginations of the plasma membrane termed *caveolae,* lined with the membrane protein *caveolin,* contain some receptor proteins and are used for certain types of receptor-mediated endocytosis. However, receptor-mediated endocytosis generally occurs via clathrin-coated pits and vesicles. In this respect, the process is similar to the packaging of lysosomal enzymes by mannose 6-phosphate (M6P) in the *trans*-Golgi. As noted earlier, although most M6P receptors are localized to the *trans*-Golgi, some are found on the cell surface.

Lysosomal enzymes that are secreted bind to these receptors and are returned to cells via receptor-mediated endocytosis. In general, transmembrane receptor proteins that are internalized from the cell surface during endocytosis are sorted and recycled back to the cell surface, much like the recycling of M6P receptors to the plasma membrane and *trans*-Golgi.

Clathrin-coated pits make up about 2 percent of the surface of cells such as hepatocytes and fibroblasts. Many internalized ligands have been observed in clathrin-coated pits and vesicles, and researchers believe that these structures function as intermediates in the endocytosis of most (though not all) ligands bound to cell-surface receptors. Some receptors are clustered over clathrin-coated pits even in the absence of ligand. Other receptors diffuse freely in the plane of the plasma membrane but undergo a conformational change when binding to ligand, so that when the receptor-ligand complex diffuses into a clathrin-coated pit, it is retained there. Two or more types of receptor-bound ligands, such as LDL and transferrin, can be seen in the same coated pit or vesicle. As discussed later, the regulated polymerization of clathrin is thought to cause the pits to expand and eventually to form clathrin-coated vesicles. Here we consider the endocytosis and subsequent sorting of cell-surface receptors and their ligands. We first describe the most common pathway exemplified by the LDL system and then briefly discuss several variations.

The LDL Receptor Binds and Internalizes Cholesterol-Containing Particles

Whether ingested in foodstuffs or synthesized in the liver, cholesterol is insoluble in body fluids and must be transported by a water-soluble carrier. Low-density lipoprotein (LDL) is one of several complexes that carry cholesterol through the bloodstream. An LDL particle is a sphere 20 – 25 nm in diameter. Its outer surface is a *monolayer* membrane of phospholipids and cholesterol, in which one molecule of a very large protein, called *apo-B*, is embedded. Inside is an extremely nonpolar core of cholesterol, all of which is esterified through the single hydroxyl group of cholesterol to a long-chain fatty acid, mainly linoleic acid. Most mammalian cells produce cell-surface receptors that specifically bind and internalize LDL by receptor-mediated endocytosis. After endocytosis, the LDL particles are transported to lysosomes where lysosomal hydrolases degrade the apo-B protein to amino acids and cleave the cholesterol esters to cholesterol and fatty acids. The cholesterol is incorporated directly into cell membranes or is reesterified and stored as lipid droplets in the cell for later use; the fatty acids are used to make new phospholipids or triglycerides. Cholesterol also is converted to steroid hormones in adrenal cortical cells and to bile acids in hepatocytes.

The LDL receptor is a single-chain glycoprotein of 839 amino acids with a long N-terminal exoplasmic domain and short C-terminal cytosolic domain. A sequence of 22 hydrophobic amino acids spans the plasma membrane once, presumably as an α helix. The large exoplasmic domain has an N-terminal segment of about 320 residues that is extremely rich in disulfide-bonded cysteine residues. This segment includes a sevenfold repeat of a sequence of 40 amino acids that contains the LDL-binding site. As explained below, residues in the C-terminal cytosolic domain are involved in trapping the LDL receptor in clathrin-coated pits.

Cytosolic Sequences in Some Cell-Surface Receptors Target Them for Endocytosis

As we learn cells possess many different cell-surface receptors with binding sites for specific ligands. Only some of these receptors, however, form receptor-ligand complexes that are internalized into clathrin-coated vesicles. Insight into what distinguishes receptors that are endocytosed from those that are not has come from studies with mutant receptors. For example, mutant forms of the LDL receptor protein are available from persons who have the inherited disorder *familial hypercholesterolemia*. This disease is characterized by high levels of cholesterol in the blood, and persons homozygous for the mutant alleles often die at an early age from heart attacks caused by atherosclerosis, a buildup of cholesterol deposits that ultimately block the arteries.

In some persons with this disorder, the LDL receptor is simply not produced; in others, it binds LDL poorly or not at all. In one especially instructive case, the mutant receptor binds LDL normally but the LDL-receptor complex cannot be internalized by the cell and is distributed evenly over the cell surface rather than confined to clathrin-coated pits. In individuals with this particular defect, plasma-membrane receptors for other ligands are internalized normally in clathrin-coated pits, but the mutant LDL receptor apparently cannot bind properly to coated pits. The mutant receptor has a single tyrosine-to-cysteine change in its cytosolic domain.

This and other mutant forms of the LDL receptor have been experimentally generated and expressed in fibroblasts. Analysis of these proteins has shown that a four-residue sequence in the cytosolic domain is crucial for internalization: Tyr-X-X-ø, where X can be any amino acid and ø is a bulky hydrophobic amino acid, such as Phe, Leu, or Met. As we discuss later, this sequence binds to one of the subunits (μ2) of the protein complex that links the clathrin coat to the cytosolic domain of a membrane protein in forming coated pits. Because the tyrosine and ø residues mediate this binding, a mutation in either one reduces or abolishes the ability of the receptor to be incorporated into clathrin-coated pits.

Many other plasma-membrane receptors that are internalized into clathrin-coated pits, such as the transferrin receptor, contain a similar amino acid sequence in their cytosolic domains. Mutagenesis studies on these receptors have confirmed that these four amino acids form a general recognition signal for binding to clathrin-coated pits.As further evidence for the importance of this sequence, a cell-surface protein that is not normally internalized into clathrin-coated pits can be made to internalize if these four amino acids are added to its cytosolic domain. For example, a mutant influenza HA protein, genetically engineered to contain such a four-amino-acid recognition sequence in its cytosolic domain, is internalized into clathrin-coated pits.

However, in other proteins different amino acids sequences, such as Leu-Leu, signal endocytosis. Yet other membrane proteins, such as the yeastα factor receptor, uracil permease, and the human growth hormone receptor require covalent addition of ubiquitin to their cytosolic domain for endocytosis to occur. At present we do not know how endocytosis of such proteins is controlled, nor the identity of the proteins that might bind to these signals.

The Acidic pH of Late Endosomes Causes Most Receptors and Ligands to Dissociate

The overall rate of endocytic internalization of the plasma membrane is quite high; cultured fibroblasts regularly internalize 50 percent of their cell-surface proteins and phospholipids each hour. Most cell-surface receptors that undergo endocytosis will repeatedly deposit their ligands within the cell and then recycle to the plasma membrane, once again to mediate the internalization of ligand molecules. For instance, the LDL receptor makes one round trip into and out of the cell every 10 – 20 minutes, for a total of several hundred trips in its 20-hour life span. In contrast, after binding its protein ligand the receptors for insulin and other growth factors generally cycle only two or three times before the complex of receptor and ligand is degraded in the lysosome — reducing the number of cell-surface receptors and thus the sensitivity of the cells to hormone signaling.

Regardless of how many times a particular receptor is recycled, internalized receptor-ligand complexes commonly follow the pathway depicted in figure. Endocytosed cell-surface receptors dissociate from their ligands within late endosomes. These acidic spherical vesicles with tubular branching membranes are found a few micrometers from the cell surface. (Similar acidic sorting vesicles recycle M6P receptors back to the Golgi complex.)

The original experiments that defined the late endosome sorting vesicle utilized the *asialoglycoprotein receptor*. This liver-specific protein mediates the binding and internalization of abnormal glycoproteins whose oligosaccharides terminate in galactose rather than the normal sialic acid, hence the name *asialo*glycoprotein. Electron microscopy of liver cells perfused with asialoglycoprotein reveal that between 5 and 10 minutes after internalization both the receptors and their ligands accumulate in late endosome vesicles. Ligand molecules are found in the lumen of the spherical part of these vesicles, while the tubular membrane extensions are rich in receptor and rarely contain ligand. Thus these membranes contain receptors that have dissociated from their ligands, indicating that the late endosome is the organelle in which receptors and ligands are uncoupled.

Beginning 15 minutes after internalization, ligands are transferred to lysosomes, but the intact receptors themselves usually are not found in these organelles. Instead, the receptor-rich elongated membrane vesicles that bud from the late endosomes mediate the recycling of receptors back to the cell surface. The spherical part of the late endosome eventually buds off transport vesicles that, with their cargo of ligand, soon fuse with lysosomes. The LDL receptor, for example, is never directed to a lysosome — to be degraded by potent lysosome proteases — until it becomes damaged in some way.

The key to why receptors release their ligands in the late endosome lies in the progressively decreasing pH encountered by internalized receptor-ligand complexes as they move through clathrin-coated vesicles and various early and late endosomes. Like the very acidic lysosomes, with an internal pH of H″4.5– 5.0, clathrin-coated vesicles and endosomes contain a V-class ATP-dependent proton pump. These vesicles also contain a Cl″ channel, allowing the proton pump to generate a significant H^+ concentration gradient, rather than the transmembrane electric potential that would form if only protons were transferred from the cytosol to the vesicle lumen. Most receptors, including the asialoglycoprotein, insulin, and LDL receptors, bind their ligands tightly at neutral pH but release their ligands if the pH is lowered to 5.0 or below. The late endosome is the first vesicle encountered by receptor-ligand complexes with a pH this low and hence is the organelle in which these and most other receptors dissociate from their tightly bound ligands.

The Endocytic Pathway Delivers Transferrin-Bound Iron to Cells

The endocytic pathway involving the transferrin receptor and its ligand differs from the LDL pathway in that the receptor-ligand complex does not dissociate in late endosomes. Nonetheless, changes in pH also mediate the sorting of receptors and ligands in the transferrin pathway, which functions to deliver iron to cells.

Transferrin, a major glycoprotein in the blood, transports iron to all tissue cells from the liver (the main site of iron storage in the body) and from the intestine (the site of iron absorption). The iron-free form, *apotransferrin,* binds two Fe^{3+} ions very tightly to form *ferrotransferrin*. All growing cells contain surface transferrin receptors that avidly bind ferrotransferrin at neutral pH, after which the receptor-bound ferrotransferrin is subjected to endocytosis. Like the components of LDL, the two bound Fe^{3+} atoms remain in the cell, but there the similarity with the fate of other endocytosed ligands, including LDL, ends: the apotransferrin part of the ligand is secreted from the cell within minutes, carried in the bloodstream to the liver or intestine, and reloaded with iron.

The explanation for the behaviour of the transferrin receptor – ligand complex lies in the unique ability of apotransferrin to remain bound to the transferrin receptor at the low pH (5.0 – 5.5) of late endosomes. At a pH of less than 6.0, the two bound Fe^{3+} atoms dissociate from ferrotransferrin and are transported from the late endosome vesicle into the cytosol (in an unknown manner). The apotransferrin formed by the dissociation of the iron atoms remains bound to the transferrin receptor and is recycled back to the surface along with the receptor. Remarkably, although apotransferrin binds tightly to its receptor at a pH of 5.0 or 6.0, it does not bind at neutral pH. Hence the bound apotransferrin dissociates from its receptor when the recycling vesicles fuse with the plasma membrane and the receptor-ligand complex encounters the neutral pH of the extracellular interstitial fluid or growth medium. The surface receptor is then free to bind another molecule of ferrotransferrin.

Some Endocytosed Proteins Remain within the Cell

In several receptor-ligand systems, found mainly in oocytes (egg cells), endocytosed material simply remains in the cells and is minimally processed. Developing insect and avian oocytes, for example, internalize yolk proteins and other proteins from the blood or surrounding cells. (Coated pits were first discovered in insect eggs, where they occupy a large portion of the plasma membrane.) A hen's egg is a single cell containing several grams of protein, virtually all of which is imported from the bloodstream by receptormediated endocytosis. Vitellogenin, a precursor of several yolk proteins, is synthesized by the liver and secreted into the bloodstream, from which it is endocytosed into the developing egg. Yolk proteins remain in storage granules within the egg and are used after fertilization as a source of amino acids and energy by the developing embryo. Egg-white proteins (e.g., ovalbumin, lysozyme, and conalbumin) that are secreted by cells lining the hen oviduct also are endocytosed by the egg cell.

Transcytosis Moves Some Ligands across Cells

As noted previously, transcytosis is used by some cells in the apical-basolateral sorting of certain membrane proteins. This process of transcellular transport, which combines endocytosis and exocytosis, also can be employed to import an extracellular ligand from one side of a cell, transport it across the cytoplasm, and secrete it from the plasma membrane at the opposite side. Transcytosis occurs mainly in sheets of polarized epithelial cells. An example of transcytosis is the movement of maternal immunoglobulins (antibodies) across the intestinal epithelial cells of the newborn mouse and human. The F_c receptor that mediates the transcytosis of immunoglobulins has the property of binding to its ligand at an acidic pH of 6 but not at neutral pH. Figure shows how a difference in the pH of the extracellular media on the two sides of intestinal epithelial cells in newborn mice allows immunoglobulins to move in one direction — from the lumen to the blood. The same process also moves maternal immunoglobulins across mammalian yolk-sac cells into the fetus.

SUMMARY

- Some extracellular ligands that bind to specific cell-surface receptors are internalized, along with their receptors, in clathrin-coated vesicles. Internalized receptor-ligand complexes undergo various fates.
- Many receptors, such as the LDL receptor, release their ligand in the acidic milieu of the late endosome; the receptors are sorted into vesicles that recycle them to the plasma membrane, while the ligands are sorted into vesicles that fuse with lysosomes. In this endocytic pathway, the released ligands are degraded by lysosomal enzymes.
- Studies with mutant LDL receptors in humans with familial hypercholesterolemia revealed a Tyr-X-X-ø signal for internalizing receptors into clathrin-coated pits. Other membrane proteins contain different endocytosis signals.
- Unlike the LDL receptor, the transferrin receptor does not release its ligand following endocytosis. Rather, the two iron atoms bound to ferrotransferrin are released in the acidic late endosome, but the resulting apotransferrin remains tightly bound to its receptor and is recycled back to the plasma membrane. At the neutral pH on the cell surface, aprotransferrin is released from the receptor.
- In several receptor-ligand systems, found mainly in oocytes, endocytosed yolk proteins and other proteins remain in the cells and are minimally processed.
- In transcytosis, endocytosed material passes all the way through the cells and is exocytosed from the plasma membrane at the opposite side. An example is the movement of maternal immunoglobulins across mammalian yolk-sac cells into the fetus and across the intestinal epithelial cells of the newborn mouse.

Chapter 6

DNA Replication, Mutation and Repair

OVERVIEW OF DNA REPLICATION

DNA replication is triggered by the expression of all required proteins, such as DNA polymerase, DNA primase, and cyclin. In yeast, the transcription factor regulating the expression of these proteins is called MCB binding factor. In mammals, the corresponding transcription factor is E2F.

Gene transcription starts from the promoter, proceeding along one direction, whereas the DNA replication starts from the replication origin, proceeding along both directions. There is only one replication origin in the genomic DNA of *E. coli* but the eukaryotic DNA contains many replication origins in each chromosome.

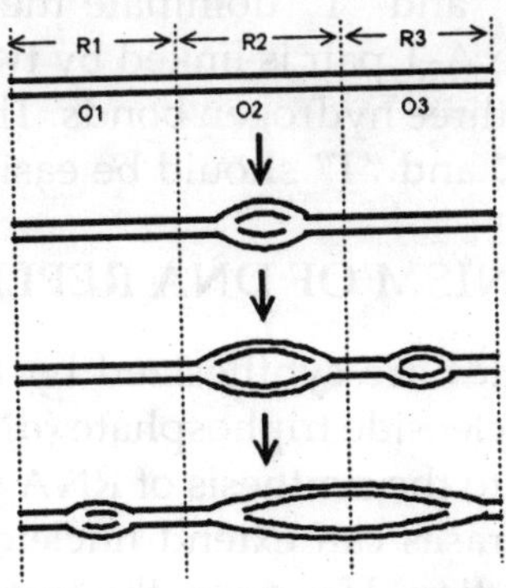

Fig. Schematic Drawing of the DNA Replication Process.

O_1, O_2, and O_3 are replication origins, each serving a region called replicon (R_1, R_2, and R_3). DNA replication involves unwinding of the double helix. New strands are synthesized by DNA polymerases using the old strands as template. Unwinding of a DNA molecule looks like a "fork" growing in one direction. The region being replicated looks like a bubble called the "replication bubble" (in red).

In *E. coli,* movement of the growing fork is about 1000 bp per second. It takes about 42 minutes to duplicate the entire genomic DNA. In eukaryotic DNA, the fork movement is only about 100 bp per second. This is probably due to the association of DNA with histones, which may hinder the fork movement. In humans, replication of the entire genome requires about 8 hours. In fruit flies, it takes only 3 - 4 minutes.

1. The bacterial replication origin contains three homologous 13-mers and four homologous 9-mers.
2. The yeast replication origin is called autonomously replicating sequence (ARS). The "A" region is absolutely necessary, while B1, B2, and B3 can increase the replication efficiency.

Based on computer analysis, the consensus sequence of human replication origin is WAWTTDDWWWDHWGW HMAWTT, where W = A or T; D = A, G or T; H = A, C or T, and M = A or C.

Note that "A" and "T" dominate the replication origin. This is because the A-T pair is linked by two hydrogen bonds while C-T pair by three hydrogen bonds. Therefore, the region dominated by "A" and "T" should be easier to unwind.

MECHANISM OF DNA REPLICATION

DNA molecules are synthesized by DNA polymerases from deoxyribonucleoside triphosphate (dNTP). The chemical reaction is similar to the synthesis of RNA strands. Both DNA and RNA polymerases can extend nucleic acid strands only in the 5' to 3' direction. However, the two strands in a DNA molecule are antiparallel. Therefore, only one strand (leading strand) can be synthesized continuously by the DNA polymerase. The other strand (lagging strand) is synthesized segment by segment (more info).

DNA POLYMERASES

E. Coli

Three types of DNA polymerases exist in *E. coli*: I, II and III. The DNA polymerase I is used to fill the gap between DNA fragments of the lagging strand. It is also the major enzyme for gap filling during DNA repair. The DNA polymerase II is encoded by the *PolB* gene, which is involved in the SOS response to DNA damage. DNA replication is mainly carried out by the DNA polymerase III.

The DNA polymerase III consists of several subunits, with a total molecular weight exceeding 600kD. Among them, α, ε, and θ subunits constitute the core polymerase. The major role of other subunits is to keep the enzyme from falling off the template strand. As shown in figure two b subunits can form a donut-shaped structure to clamp a DNA molecule in its centre, and slide with the core polymerase along the DNA molecule. This allows continuous polymerization of up to 5×10^5 nucleotides. In the absence of β subunits, the core polymerase would fall off the template strand after synthesizing 10-50 nucleotides.

Mammals

There are five types of DNA polymerases in mammalian cells: α, β, γ, δ, and ε. The γ subunit is located in the mitochondria, responsible for the replication of mtDNA. Other subunits are located in the nucleus. Their major roles are given below:

- α: synthesis of lagging strand.
- β: DNA repair.
- γ: synthesis of leading strand.
- δ: DNA repair.

SYNTHESIS OF DNA

There is a major difference between DNA polymerase and RNA polymerase: the RNA polymerase can synthesize a new strand whereas the DNA polymerase can only extend an existing strand. Therefore, to synthesize a DNA molecule, a short RNA molecule (~ 5 – 12 nucleotides) must be synthesize first by a special enzyme. The initiating RNA molecule is known as a primer, and the enzyme is called primase.

In addition to DNA polymerase and primase, DNA replication requires helicase and single strand binding protein (SSB protein). The role of helicase is to unwind the duplex DNA. SSB proteins can bind to both separated strands, preventing them from annealing (reconstitution of double-stranded DNA from single strands).

The replication mechanisms in both bacteria and eukaryotes are similar. However, eukaryotic DNA polymerases do not contain α subunit similar to the *E. coli* b subunit. They use a separate protein called proliferating cell nuclear antigen (PCNA) to clamp the DNA.

DNA polymerases can extend nucleic acid strands only in the 5' to 3' direction. However, in the direction of a growing fork, only one strand is from 5' to 3'. This strand (the leading strand) can be synthesized continuously. The other strand (the lagging strand), whose 5' to 3' direction is opposite to the movement of a growing fork, should be synthesized discontinuously.

1. Comparison between the leading strand and the lagging strand.
2. The primase first synthesizes a new primer which is about 10 nucleotides in length. The distance between two primers is about 1000-2000 nucleotides in bacteria, and about 100-200 nucleotides in eukaryotic cells.
3. DNA polymerase elongates the new primer in the 5' to 3' direction until it reaches the 5' end of a neighboring primer. The newly synthesized DNA is called an Okazaki fragment.
4. In *E. coli,* DNA polymerase I has the 5' to 3' exonuclease activity, which is used to remove a primer.

5. DNA ligase joins adjacent Okazaki fragments.

The whole lagging strand is synthesized by repeating steps (b) to (e).

TELOMERASE AND AGING

Synthesis of the lagging strand requires a short primer which will be removed. At the extreme end of a chromosome, there is no way to synthesize this region when the last primer is removed. Therefore, the lagging strand is always shorter than its template by at least the length of the primer. This is the so-called "end-replication problem".

Bacteria do not have the end-replication problem, because its DNA is circular. In eukaryotes, the chromosome ends are called telomeres which have at least two functions:

- To protect chromosomes from fusing with each other.
- To solve the end-replication problem.

The procedure to solve the end-replication problem is outlined in figure. Mechanism of the telomere extension by telomerase is explained in figure.

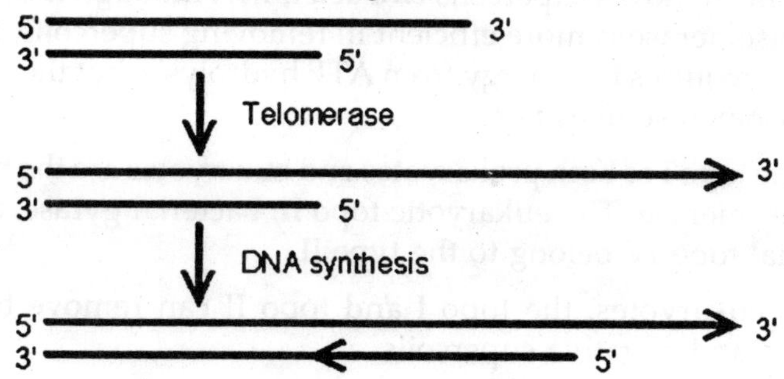

Fig. Telomerase and Telomere Extension

To extend the length of a telomere, the telomerase first extends its longer strand. Then, using the same mechanism as synthesizing the lagging strand, the shorter strand is extended.

In a human chromosome, the telomere is about 10 to 15 kb in length, composed of the tandem repeat sequence: TTAGGG. The telomerase (web link) contains an essential

RNA component which is complementary to the telomere repeat sequence. Hence, the internal RNA can serve as the template for synthesizing DNA. Through telomerase translocation, a telomere may be extended by many repeats.

AGING

In the absence of telomerase, the telomere will become shorter after each cell division. When it reaches a certain length, the cell may cease to divide and die. Therefore, telomerase plays a critical role in the aging process.

TOPOISOMERASES

During replication, the unwinding of DNA may cause the formation of tangling structures, such as supercoils or catenanes. The major role of topoisomerases is to prevent DNA tangling.

There are two types of topoisomerases: type I produces transient single-strand breaks in DNA and types II produces transient double-strand breaks. As a result, the type I enzyme removes supercoils from DNA one at a time, whereas the type II enzyme removes supercoils two at a time. Although the type II topoisomerase is more efficient in removing supercoils, this enzyme requires the energy from ATP hydrolysis, but the type I topoisomerase does not.

The topo I of both prokaryotes and eukaryotes are the type I topoisomerase. The eukaryotic topo II, bacterial gyrase, and bacterial topo IV belong to the type II.

In eukaryotes, the topo I and topo II can remove both positive and negative supercoils.

In bacteria, the topo I can remove only negative supercoils. The bacterial topo II is also called the gyrase, which has two functions:

a. To remove the positive supercoils during DNA replication,

b. To introduce negative supercoils (one supercoil for 15-20 turns of the DNA helix) so that the DNA molecule can be packed into the cell. During replication, these negative supercoils are removed by topo I.

The bacterial topo IV belongs to the type II topoisomerase. This enzyme is involved in decatenation.

1. To remove supercoils. This involves a double-strand break (indicated by a short line), allowing the tangled segment to pass through. The break is then resealed.
2. To remove catenanes. The topo II makes a double-strand break in one DNA molecule (the blue one), allowing the other molecule to pass through. The break is then resealed.

Without topoisomerases, the DNA cannot replicate normally. Therefore, the inhibitors of topoisomerases have been used as anti-cancer drugs to stop the proliferation of malignant cells. However, these inhibitors may also stop the division of normal cells. Some cells (e.g., hair cells) which need to continuously divide will be most affected. This explains a noticeable side effect: the hair loss.

MUTATIONS AND CONSEQUENCES

Mutation refers to the change in a DNA sequence, which may involve only a few bases or the large-scale chromosome abnormality. This section covers the small-scale mutations (substitution, deletion, insertion) and the exon skipping that results from mutation at the splice site.

SUBSTITUTION

In the substitution mutation, one or more nucleotides are substituted by the same number of different nucleotides. In most cases, only one nucleotide is changed. Based on the change in the nucleotide type, the substitution mutation may be divided into transition and transversion mutations. Based on the consequence of mutation, the substitution mutation may be grouped into silent, missense and nonsense mutations.

1. Illustration of transition (blue) and transversion (red) mutations. In the transition mutation, a pyrimidine

(C or T) is substituted by another pyrimidine, or a purine (A or G) is substituted by another purine. The transversion mutation involves the change from a pyrimidine to a purine, or vice versa.

2. Examples of silent, missense and nonsense mutations. The silent mutation does not produce any change in the amino acid sequence, the missense mutation results in a different amino acid, and the nonsense mutation generates a stop signal.

DELETION

The deletion mutation involves elimination of one or more nucleotides from a DNA sequence. It may cause frameshift, producing a non-functional protein.

1. Deletion of "T" from the sequence "TTTTT" in the *CFTR* gene.
2. Deletion of "AT" from the sequence "ATAT" in the *CFTR* gene.
3. Deletion of "TTG" from the sequence "TTGTTG" in the *FIX* gene.
4. Deletion of "ATAG" from the sequence "ATAGATAG" in the *APC* gene.

Note that deletion and insertion mutations often occur in the repetitive sequence. As explained in the next section, they are usually caused by "replication slippage".

INSERTION

In the insertion mutation, one or more nucleotides are inserted into a sequence. If the number of inserted bases is not a multiple of 3, it will cause frameshift, resulting in serious consequences. As shown in the following table, non-frameshift insertions may also cause diseases.

EXON SKIPPING

Splicing of an intron requires an essential signal: "GT........AG". If the splice acceptor site AG is mutated the splicing machinery will look for the next acceptor site. As a result, the exon between two introns is also removed.

MUTATION MECHANISMS

Mutations may be caused by external factors (UV light, chemical agents, etc.) or spontaneous cellular processes (replication errors, accidental deamination, etc.). The following figure illustrates the deamination process. Conversion of the methylated cytosine (methylcytosine) to thymine may explain the observed CpG islands (more info).

Cytosine is one of four bases in DNA molecules. As shown in the above figure, it may be mutated to uracil by deamination. Since uracil is not part of DNA, this mutation can easily be detected and repaired by base excision. Suppose DNA, like RNA, were made up of uracil, then the cytosine to uracil mutation could be corrected only by mismatch repair which is very inefficient. This may explain why DNA chooses thymine, instead of uracil, even though the chemical structure of uracil is simpler than thymine.

MUTATION BY UV LIGHT

UV light may cause two adjacent pyrimidine residues (cytosine or thymine) to form a dimer. In a normal cell, the dimer can be detected by p53, which then triggers the repairing process. However, if p53 itself is mutated and become non-functional, the pyrimidine dimer may lead to mutation.

In the above figure, the UV light first causes two adjacent cytosine residues to form a dimer. During DNA replication, both strands are used as templates to synthesize new strands. The cytosine dimer could cause adenine (instead of the normal guanine) to be incorporated into the new strand. Subsequent DNA replication will produce CC to TT mutation. Although the cytosine dimer may eventually be corrected, the mutation cannot be detected by the DNA repair system.

MUTATION BY CHEMICAL AGENTS

The chemical agents which may cause mutation are called mutagens. Most of them are also carcinogens. The following figure shows the chemical structures of a few potent mutagens. Figure explains the mechanism of mutation by 5-bromouracil.

- Acridines (e.g., proflavin) are positively charged molecules. They may be inserted between two DNA strands, thereby altering DNA's structure and rigidity. As a result, DNA replication will not be faithful.
- Alkylating agents are chemicals that add an alkyl group (C_nH_{2n+1}) to another molecule. Alkylation of a base may change the normal base pairing. For example, the alkylating agent EMS converts guanine to 7-ethylguanine which pairs with thymine. The mispairing will lead to mutation. Some alkylating agents may also cross-link DNA, resulting in chromosome breaks.
- Nitrous acid is a deaminating agent that converts cytosine to uracil, adenine to hypoxanthine, and guanine to xanthine. The hydrogen-bonding potential of the modified base is altered, resulting in mispairing.
- Hydroxylamine and free radicals may modify base structures, resulting in mispairing.

This molecule has two tautomeric isoforms. Its keto form (BU_k) pairs with adenine whereas its enol form (BU_e) pairs with guanine. Suppose in the first replication the keto form was incorporated into a new DNA strand. During the second replication, if the keto form undergoes a tautomeric shift to the enol form, it will cause A:T to G:C mutation.

MUTATION BY REPLICATION ERRORS

Replication errors are the main source of mutations. It has been estimated that uncorrected replication errors occur with a frequency of 10^{-9} - 10^{-11} for each nucleotide added by DNA polymerases. Since a cell division requires synthesis of 6 X 10^9 nucleotides, the mutation rate is about one per cell division.

A commonly observed replication error is the replication slippage, which occurs at the repetitive sequences when the new strand mispairs with the template strand. The microsatellite polymorphism is mainly caused by the replication slippage. If the mutation occurs in a coding region, it could produce abnormal proteins, leading to diseases. The Huntington's disease is a well known example.

DNA METHYLATION AND CPG ISLAND

Origin of CpG (CG) islands

The CG island is a short stretch of DNA in which the frequency of the CG sequence is higher than other regions. It is also called the CpG island, where "p" simply indicates that "C" and "G" are connected by a phosphodiester bond.

CpG islands are often located around the promoters of housekeeping genes (which are essential for general cell functions) or other genes frequently expressed in a cell. At these locations, the CG sequence is not methylated. By contrast, the CG sequences in inactive genes are usually methylated to suppress their expression. The methylated cytosine may be converted to thymine by accidental deamination. Unlike the cytosine to uracil mutation which is efficiently repaired, the cytosine to thymine mutation can be corrected only by the mismatch repair which is very inefficient. Hence, over evolutionary time scales, the methylated CG sequence will be converted to the TG sequence. This explains the deficiency of the CG sequence in inactive genes.

Inheritance of the DNA methylation pattern

Any type of cells have their own methylation pattern so that a unique set of proteins may be expressed to perform functions specific for this cell type. Thus, during cell division, the methylation pattern should also pass over to the daughter cell. This is achieved by a specific enzyme called the maintenance methylase which can methylate only the CG sequence paired with methylated CG.

DNA REPAIR MECHANISMS

There are three major DNA repairing mechanisms: base excision, nucleotide excision and mismatch repair.

Base Excision

DNA's bases may be modified by deamination or alkylation. The position of the modified (damaged) base is called the "abasic site" or "AP site". In *E.coli*, the DNA glycosylase can recognize the AP site and remove its base. Then, the AP endonuclease removes the AP site and neighboring nucleotides. The gap is filled by DNA polymerase I and DNA ligase.

Nucleotide Excision

In *E. coli*, proteins UvrA, UvrB, and UvrC are involved in removing the damaged nucleotides (e.g., the dimer induced by UV light). The gap is then filled by DNA polymerase I and DNA ligase. In yeast, the proteins similar to Uvr's are named RADxx ("RAD" stands for "radiation"), such as RAD3, RAD10. etc.

Mismatch Repair

To repair mismatched bases, the system has to know which base is the correct one. In *E. coli*, this is achieved by a special methylase called the "Dam methylase", which can methylate all adenines that occur within (5')GATC sequences. Immediately after DNA replication, the template strand has been methylated, but the newly synthesized strand is not methylated yet. Thus, the template strand and the new strand can be distinguished.

The repairing process begins with the protein MutS which binds to mismatched base pairs. Then, MutL is recruited to the complex and activates MutH which binds to GATC sequences. Activation of MutH cleaves the unmethylated strand at the GATC site. Subsequently, the segment from the cleavage site to the mismatch is removed by exonuclease (with assistance from helicase II and SSB proteins). If the cleavage occurs on the 3' side of the mismatch, this step is carried out by exonuclease I (which degrades a single strand only in the 3' to 5' direction). If the cleavage occurs on the 5' side of the mismatch, exonuclease VII or RecJ is used to degrade the single stranded DNA. The gap is filled by DNA polymerase III and DNA ligase.

The distance between the GATC site and the mismatch could be as long as 1,000 base pairs. Therefore, mismatch repair is very expensive and inefficient.

Mismatch repair in eukaryotes may be similar to that in *E. coli*. Homologs of MutS and MutL have been identified in yeast, mammals, and other eukaryotes. MSH1 to MSH5 are homologous to MutS; MLH1, PMS1 and PMS2 are homologous to MutL. Mutations of MSH2, PMS1 and PMS2 are related to colon cancer.

In eukaryotes, the mechanism to distinguish the template strand from the new strand is still unclear.

Chapter 7

Cell Division and System Development

CELL CYCLE

To understand the molecular mechanism of cell division, we need to know the life cycle of a cell. It consists of four phases: G_1, S, G_2 and M, where "G" stands for "Gap", "S" represents "Synthesis" and "M" means "Mitosis". A newly divided cell may either enter into another round of cell division or remain in the resting state for a long period. In the latter case, the cell is said to be in the G_0 phase. Upon specific stimulation, cells in the G_0 phase may re-enter the G_1 phase.

CDK and Cyclins

The progression of cell cycle is catalyzed by cyclin-dependent kinase (CDK) which, as the name suggests, is activated by a special class of proteins called cyclins. In mammals, different cyclins are designated as A, B, C, D and so on. In budding yeast nine cyclins have been identified: Cln1 to Cln3 and Clb1 to Clb6.

The proteins of the CDK superfamily in mammals is denoted by Cdk followed by a number. However, in yeast, they are denoted by Cdc (cell division control) and a number. Among them, Cdc2 was first identified in fission yeast (*S. pombe*). It is equivalent to Cdc28 in budding yeast, and to Cdk1 in mammals. Cdk1 is often referred to as Cdc2.

In budding yeast, the cell cycle is initiated by the binding of Cln1 or Cln2 to Cdc28. Then, Clb5 or Clb6 binds to Cdc28, and pushes the cycle forward. Clb1 and Clb2 are involved in the final phase.

In mammals, the cell cycle is initiated by the binding of Cyclin D (CycD) to Cdk4 or Cdk6. The final phase is catalyzed by the binding between Cyclin B (CycB) and Cdk1. The CycB-Cdk1 complex is known as MPF, which originally stands for maturation promoting factor, but now also stands for mitosis promoting factor.

INITIATION AND TERMINATION MECHANISM OF S PHASE

DNA replication occurs in the S phase of the cell cycle. DNA replication is triggered by the expression of all required proteins. In mammals, the expression of these necessary proteins is activated by the transcription factor E2F which, in turn, is regulated by the protein pRB (the protein product of the retinoblastoma gene). When pRB binds to E2F, it will inactivate the transcriptional function of E2F, thereby inhibiting DNA replication. With such a critical role, pRB is known as the "master brake of cell division".

During G_0 or early G_1 phase, E2F is inactivated by pRB. In late G_1 phase, the Cyclin D-Cdk4 complex phosphorylates a specific set of sites on pRB, resulting in the release of E2F, which then leads to the production of proteins needed for DNA replication. E2F also stimulates the production of Cyclin E, cyclin A and E2F itself. The Cyclin E-Cdk2 complex can also phosphorylate pRB, thus accelerating the cell cycle advance into the S phase. On the other hand, the Cyclin A-Cdk2 complex can phosphorylate E2F, abolishing its transcriptional function. Consequently, DNA replication is terminated.

It is interesting to note that the Cyclin E-Cdk2 complex cannot phosphorylate E2F. Otherwise, the DNA replication would be terminated prematurely.

MITOSIS

Mitosis is a process of cell division resulting in two genetically equivalent daughter cells. The period between mitoses is called the interphase which covers the G_1, S, and G_2 phases of the cell cycle. During the G_2 phase, DNA has been replicated and the chromosome number is 4n. In the interphase, chromosomes are not visible by light microscopy because the chromatin diffuses throughout the nucleus.

Prophase: The chromatin condenses into chromosomes. Two centrioles move toward the opposite ends (poles) of the cell. Microtubules radiate from each centriole, forming a structure called spindle. By the end of prophase, the nuclear envelope breaks down into small vesicles.

Metaphase: The chromosomes align at the equatorial plane of the cell, as directed by the microtubules attached to the kinetochore (located at the centromere of each chromosome).

Anaphase: Sister chromatids separate and move apart. The separation happens simultaneously for the entire set of chromosomes. The microtubules attached to the kinetochore also play a critical role in this process.

Telophase: The chromosomes begin to uncoil and become less condensed. The spindle disappears while the nuclear envelope reappears. The cell elongates and eventually divides into two daughter cells. The process of cell division is called cytokinesis

The morphological changes in early mitosis are mainly due to the phosphorylation of related proteins catalyzed by MPF (more info). The mitosis is terminated by the activation of DBRP.

MPF AND THE INITIATION OF MITOSIS

The morphological changes in early mitosis are mainly due to the phosphorylation of related proteins catalyzed by MPF (mitosis promoting factor), which is actually the Cyclin B/Cdk1 complex. The activated MPF may phosphorylate histone protein H1 to condense chromatin. It can also phosphorylate nuclear lamins to break nuclear envelope. The microtubular structure is closely related to MAPs (microtubule-associated proteins) which can also be phosphorylated by MPF, resulting in the formation of spindle.

The activity of MPF depends on its phosphorylation state, as shown in the following figure.

TERMINATION MECHANISM OF MITOSIS

The mitosis is terminated by activated DBRP (destruction box recognition proteins) which can degrade Cyclin B and Cyclin A. DBRP is activated by phosphorylation catalyzed by MPF (the Cyclin B/Cdk1 complex). During mitosis, the activity of MPF keeps rising until it phosphorylates DBRP to degrade Cyclin B, thereby inhibiting MPF. As explained previously, the morphological changes in early mitosis are mainly due to the phosphorylation of related proteins by MPF. After MPF is inhibited in late mitosis, those morphological changes are reversed: chromatin less condensed, nuclear envelope reformed and spindle disappears.

MEIOSIS

Meiosis is a special type of cell division resulting in four genetically nonequivalent daughter cells, each containing half the number of chromosomes of the parent cell. Germ cells (eggs and sperm) are produced by meiosis.

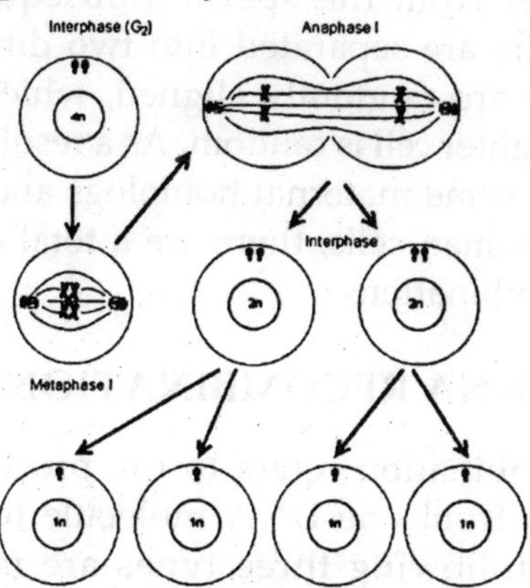

Fig. Illustration of Meiosis.

The meiosis involves two consecutive cell divisions. Each division is composed of four stages as in mitosis. However, the first division differs in two important aspects:

1. In the metaphase I, each pair of sister chromatids aligns with its homologous pair side by side, forming two lines of sister chromatids at the equatorial plane (there is only one line in mitosis). Such alignment is called synapsis.
2. In the subsequent anaphase I, each pair of sister chromatids moves away from its homologous pair, but the sister chromatids are not separated.

After the first division, the chromosome number in each daughter cell becomes 2n. The process of the second division is the same as in mitosis except that there is no DNA replication during the interphase. As a result, the produced cell is a haploid (1n).

During synapsis, DNA recombination may occur between homologous pairs of sister chromatids.

INDEPENDENT ASSORTMENT

In the metaphase I of meiosis, two pairs of homologous sister chromatids align side by side. One pair (the maternal homolog) comes from the egg and another pair (the maternal homolog) comes from the sperm. Subsequently, the two homologous pairs are separated into two different daughter cells. Since they are randomly aligned, which homolog will enter which daughter cell is random. As a result, each daughter cell will contain some maternal homologs and some paternal homologs. For human cells, there are a total of 8.4 million (= 2^{23}) possible combinations.

DNA RECOMBINATION

DNA recombination refers to the process that a DNA segment moves from one DNA molecule to another DNA molecule. The following three types are most commonly observed.

HOMOLOGOUS RECOMBINATION

Homologous recombination occurs between two homologous DNA molecules. It is also called DNA crossover. During meiosis, two homologous pairs of sister chromatids align side by side. The DNA crossover is very likely to occur. It could be as often as several times per meiosis.

The mechanism of DNA crossover was first explained by Robin Holliday in 1964 (More Info).

Due to significant sequence homology, the repetitive sequence region of a chromatid may not line up exactly with its corresponding region in a homologous chromatid or identical sister chromatid. This may result in unequal crossover, which is thought to be the major mechanism underlying VNTR (variable number of tandem repeats).

DNA crossover is not the same as gene conversion.

SITE-SPECIFIC RECOMBINATION

Site-specific recombination occurs at a specific DNA sequence. The first example was found in the integration between l DNA and *E. coli* DNA. Both of them contain a sequence, 5'-TTTATAC-3', called the attachment site, which allows the two DNA molecules to attach together by base pairing. Once attached, the enzyme integrase catalyzes two single strand breaks as in the Holliday model. After a short branch migration, the integrase exerts a second strand cuts on two other strands. Resolution of two Holliday junctions completes the integration process.

TRANSPOSITIONAL RECOMBINATION

Transpositional recombination is a process in which a mobile element is inserted into a target DNA. It may occur by one of two mechanisms:

1. Directly as DNA,
2. Through RNA. The mobile elements that transpose through DNA are called transposons and those via RNA are referred to as retrotransposons. Some examples are given in the following figure.

Mechanisms of Transpositional Recombination

a. For transposons. The transposon in donor DNA is cut by a special enzyme and then inserted into a target DNA. In bacteria, this enzyme is called transposase which has both nuclease and ligase activities.

b. For retrotransposons. The retrotransposon in donor DNA is first transcribed into RNA and then reverse-transcribed into DNA, which is inserted into a target DNA by the same recombination mechanism as the DNA intermediates of transposons.

The transpositional recombination as shown above involves DNA recombination whose mechanism is explained in the following figure.

In bacteria, the target DNA is cut by transposase, producing sticky ends (single strands, easy to pair with complementary sequences). The transposase also has ligase activity which may ligate the intermediate DNA of transposons. The gaps at sticky ends are filled by DNA polymerase, generating direct repeats.

Other organisms are expected to use the same mechanism to insert the DNA intermediates of either transposons or retrotransposons into target DNA. The direct repeats generated by this mechanism is found in ALL mobile elements.

In recombinant DNA technology, a DNA fragment is inserted into a target DNA by a slightly different mechanism.

The homologous recombination often occurs during meiosis. Other types of recombination are not specifically related to cell division.

THE HOLLIDAY MODEL OF DNA CROSSOVER

The widely accepted model for DNA crossover was first proposed by Robin Holliday in 1964. It involves several steps as illustrated in the following figure.

1. Two homologous DNA molecules line up (e.g., two nonsister chromatids line up during meiosis).
2. Cuts in one strand of both DNAs.

3. The cut strands cross and join homologous strands, forming the *Holliday structure* (or *Holliday junction*).
4. Heteroduplex region is formed by *branch migration.*
5. Resolution of the Holliday structure. DNA strands may be cut along either the vertical line or horizontal line.
6. The vertical cut will result in crossover between f-f′ and F-F′ regions. The heteroduplex region will eventually be corrected by mismatch repair.
7. The horizontal cut does not lead to crossover after mismatch repair. However, it could cause gene conversion.

The detailed mechanism of homologous recombination was mainly obtained from the study of *E. coli*. Although bacteria do no undergo meiosis, homologous recombination could occur during or immediately after DNA replication. It may also occur in a mating process called conjugation.

In *E. coli,* the recombination is initiated by the enzyme RecBCD, consisting of three subunits: RecB, RecC and RecD. This enzyme has both helicase and nuclease activities. The enzyme first uses its helicase activity to unwind DNA.When it hits the Chi site (with sequence GCTGGTGG), one of the exposed strand will be cut by its nuclease activity. The reason why this special site is called the "Chi site" is because the Greek letter c (chi) looks like a crossover point. The Chi site is the position of the Holliday junction and also the position of chiasma.

After DNA strands are cut by RecBCD, the strand invasion is catalyzed by RecA proteins, which can wrap around single stranded DNA and direct it to form the Holliday structure.

Finally, the branch migration is catalyzed by RuvA and RuvB. The Holliday structure is resolved by the protein RuvC.

The original Holliday model involves two single-strand breaks. Evidence suggests that the Holliday structure could also be initiated by other alternatives:

- Double-strand break: From Kimball's Biology Pages.
- One single-strand break: Proposed by M. Meselson and C. Radding in 1975 (Proc. Natl. Acad. Sci. USA, 72:358).

GENE CONVERSION

Gene conversion is a nonreciprocal transfer of genetic information. The following figure illustrates the difference between gene conversion and DNA crossover. An origin of gene conversion is explanined in figure.

UNEQUAL CROSSOVER

Due to significant sequence homology, the repetitive sequence region of a chromatid may not line up exactly with its corresponding region in a homologous chromatid or identical sister chromatid. As a result, different number of repeat units may be generated during meiosis. This is thought to be the major mechanism underlying VNTR (variable number of tandem repeats).

1. Two pairs of sister chromatids line up during meiosis. A repetitive region of one chromatid (the third one) does not line up exactly with its corresponding region in other chromatids.
2. Strand breaks on nonsister chromatids (along line A) will result in unequal crossover, producing different number of repeat units in these chromatids.

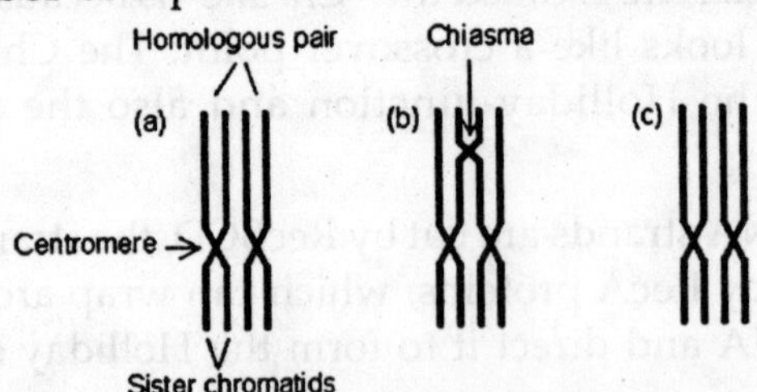

3. Strand breaks on sister chromatids (along line B) also produce different repeats. In this case, it is called sister chromatid exchange. The detailed mechanism of DNA crossover in (b) and (c) may be explained by the Holliday model.

SITE-SPECIFIC RECOMBINATION

Site-specific recombination occurs at a specific DNA sequence. The first example was found in the integration between l DNA and *E. coli* DNA. Both of them contain a sequence, 5'-TTTATAC-3', called the attachment site, which allows the two DNA molecules to attach together by base pairing. Once attached, the enzyme integrase catalyzes two single strand breaks as in the Holliday model. After a short branch migration, the integrase exerts a second strand cuts on two other strands. Resolution of two Holliday junctions completes the integration process.

CHROMOSOME ABNORMALITY

NUMERICAL ABNORMALITY

Triploidy

The cell contains three haploid sets of chromosomes (69,XXX or 69,XXY) caused by one of the following mechanisms:

1. Fertilization of a single egg by two sperm.
2. Fertilization between a normal egg and an abnormal diploid sperm.
3. Fertilization between an abnormal diploid egg and a normal sperm.

Monosomy

The cell has only one copy of a specific chromosome. For example, individuals with Turner's syndrome have a single X chromosome and no Y.

Trisomy

The cell has an extra copy of a specific chromosome. Examples:

- Down's syndrome - with three copies of chromosome 21.
- Klinefelter's syndrome - with three sex chromosomes (XXY).

Both monosomy and trisomy arise from the same mechanism as explained in the following figure.

a. During mitotic metaphase or meiotic metaphase II, one pair of sister chromatids does not align at the equatorial plane. Consequently, both sister chromatids move to the same cell. One daughter cell will show trisomy and another daughter cell will have monosomy.

b. During meiotic metaphase I, two homologous pairs of sister chromatids do not align at the equatorial plane and subsequently move to the same cell (a phenomenon called nondisjunction). The resulting daughter cells also have either monosomy or trisomy.

STRUCTURAL ABNORMALITY

Chromosomal Translocation

Chromosomal translocation is the interchange of parts between nonhomologous chromosomes. It is caused by breaks in more than one chromosome, followed by attachment of a chromosome fragment to a different chromosome. There are two main types: reciprocal translocation and Robertsonian translocation.

Breaks occur first in both chromosome 4 and chromosome 20. Then one fragment of chromosome 4 attaches to a fragment of chromosome 20 and another fragment of chromosome 4 attaches to another fragment of chromosome 20. Picture courtesy: National Human Genome Research Institute

Robertsonian translocation is caused by breaks at the centromeres of two nonhomologous acrocentric chromosomes (chromosomes with centromeres near their ends, e.g., chromosome 13, 14, 21, and 22). After the breaks, the long arms of two chromosomes join together to form a single chromosome. The short arms may join to form another chromosome, but it is usually lost within a few cell divisions. Therefore, the karyotype of a person with Robertsonian translocation usually exhibits only 45 chromosomes.

In addition to the trisomy 21 mentioned above, the Down's syndrome may also be caused by Robertsonian translocation, in which the long arm of chromosome 21 is attached to another chromosome, usually chromosome 14 or itself.

Inversion, Deletion and Ring Structure

Two breaks in a single chromosome can cause inversion, deletion or ring structure as shown in the following figure.

MECHANISMS OF DEVELOPMENT

GENE CONTROL IN DEVELOPMENT

The ultimate goal of developmental biologists is to unravel the mystery of how a fertilized egg is transformed into a complex multicellular organism. This process requires execution of a complex developmental programme whereby specific genes are activated in a precise time sequence and in the correct location, generating different types of tissues and the specific cell types composing them. Classical and molecular geneticists have discovered numerous genes that participate in the highly regulated programmes that result in the development of plants and animals. Understanding the molecular basis for the action of such genes is one of the most actively studied areas in all of biology.

We examined various mechanisms for regulating gene expression. As a result of these controls, cells can respond to changes in their environment and different cell types can produce characteristic proteins. By far the most prominent mechanism for regulating the production of different proteins in different cells entails a vast array of DNA-binding proteins that act, often in various combinations, to either activate or repress gene transcription.

In this chapter, we take a closer look at the spatial and temporal control of gene expression during development. Genetic and molecular studies show that different cells express different sets of genes based on their developmental history, their patterns of cell division, their position in the developing organism, and their interactions with other cells. We focus on several well-studied cases of differential gene transcription to specify different cell types in yeast, animals, and plants to illustrate the general transcription-control strategies that regulate development.

CELL INTERACTIONS IN DEVELOPMENT

We learned that regionalization along the anteroposterior axis in the early *Drosophila* embryo is largely determined by gradients of transcription factors generated through translation of spatially restricted maternal mRNAs and subsequent diffusion of the encoded proteins through the common cytoplasm of the syncytial blastoderm. These transcription factors, in turn, control the patterned expression of specific target genes along the anterioposterior axis. In contrast, local interactions between cells, mediated by secreted or cell-surface signaling molecules, determine regionalization along the dorsoventral axis in *Drosophila* and along both major axes in early vertebrate embryos. Such local interactions also are the primary mechanism regulating the formation of internal organs such as the kidney, lung, and pancreas. Likewise, the vast number of highly specialized cells and their stereotyped arrangement in different tissues is a consequence of locally acting signals.

The importance of cellular interactions in development was demonstrated first in the early part of twentieth century through two complementary experiments. In one, destruction of an optic-vesicle primordium in developing frogs prevented formation of the lens from the overlying ectodermal cells. Conversely, transplantation of an optic-vesicle primordium to a region of ectoderm that normally does not give rise to a lens induced formation of a lens in an abnormal *(ectopic)* site. In modern biology we now use the term induction to refer to any mechanism whereby one cell population influences the development of neighboring cells.

In some cases, induction involves a binary choice. In the presence of a signal the cell is directed down one developmental pathway; in the absence of the signal, the cell assumes a different developmental fate or fails to develop at all. In other cases, signals can induce different responses in cells at different concentrations. For instance, a low concentration of an inductive signal causes a cell to assume fate A, but a higher concentration causes the cell to assume fate B. The concentration at which a signal induces a specific cellular response is called a *threshold*.

In many cases, an inductive signal induces an entire tissue containing multiple cell types. Two models have been proposed to account for these properties of extracellular signaling molecules. In the *gradient model*, a signaling molecule induces different fates at different threshold concentrations. A cell's fate, then, is determined by its distance from the signal source. In the alternative *relay model*, a signal induces a cascade of induction in which cells close to the signal source are induced to assume specific fates; they, in turn, produce other inductive signals to pattern their neighbors.

Although inductive interactions often are unidirectional, they sometimes are reciprocal. Prominent examples of reciprocal induction include the formation of internal organs such as the kidney, pancreas, and lung. Many inductive interactions occur between *non-equivalent* cells; that is, the signaling and responding cells are already different. However, interactions between *equivalent* cells often are crucial in assuring that some cells in a developing tissue assume a specific fate and others do not. An evolutionarily conserved class of ligands and receptors regulates such interactions in *C. elegans, Drosophila,* and vertebrates.

Another feature that distinguishes various developmental pathways is the nature of the extracellular inductive signals. Many are freely diffusible and hence can act at a distance, whereas some are tethered to the cell surface and are available only to immediate neighboring cells. Still others are highly localized by their tight binding to the extracellular matrix. Early embryologists noted that cells differed in their ability to respond to inducing signals. Cells that can respond to such signals are referred to as *competent*. Competence may reflect the expression of receptors specific for a given signaling molecule, the ability of the receptors to activate specific intracellular signaling pathways, or the presence of the transcription factors necessary to stimulate expression of the genes required to implement the developmental programme induced.

In this chapter, we first describe examples of various types of inductive signals and cellular interactions that regulate cell-type specification in several different developmental systems. Specific extracellular signals also control the migration of certain cells, which occurs during development of some tissues. As an example of this phenomenon, we discuss the role of extracellular signals in the assembly of connections between neurons. Another common feature of developmental programmes is the highly regulated death of certain cells. In the final section of this chapter, we examine the conserved pathway leading to cell death and how it is controlled. The examples presented in this chapter were chosen to illustrate key concepts in this rapidly advancing field.

STEM CELLS

Stem cells have three important characteristics that distinguish them from other types of cells.

- They are capable of dividing and renewing themselves for long periods.
- They are unspecialized.
- Under certain physiologic or experimental conditions, they can be induced to become specialized cells such as nerve cells or blood cells.

Stem cells represent natural units of embryonic development and tissue regeneration. Embryonic stem (ES) cells, in particular, possess a nearly unlimited self-renewal capacity and developmental potential to differentiate into virtually any cell type of an organism. Mouse ES cells, which are established as permanent cell lines from early embryos, can be regarded as a versatile biological system that has led to major advances in cell and developmental biology. Human ES cell lines, which have recently been derived, may additionally serve as an unlimited source of cells for regenerative medicine. Before therapeutic applications can be realized, important problems must be resolved. Ethical issues surround the derivation of human ES cells from in vitro fertilized blastocysts. Current techniques for directed differentiation into somatic cell populations remain inefficient and yield heterogeneous cell populations. Transplanted ES cell progeny may not function normally in organs, might retain tumorigenic potential, and could be rejected immunologically. The number of human ES cell lines available for research may also be insufficient to adequately determine their therapeutic potential. Recent molecular and cellular advances with mouse ES cells, however, portend the successful use of these cells in therapeutics. This review therefore focuses both on mouse and human ES cells with respect to in vitro propagation and differentiation as well as their use in basic cell and developmental biology and toxicology and presents prospects for human ES cells in tissue regeneration and transplantation.

FORMATION OF TISSUES

INTEGRATING CELLS INTO TISSUES

The evolution of multicellular organisms permitted specialized cells and tissues to form; a flowering plant has at least 15 cell types, and a vertebrate hundreds. In both plants and animals, cells that are specialized to carry out a particular task are found together in the tissues in which the task is performed: a xylem or meristem; a liver, a muscle, or a nerve ganglion. Different types of cells in a tissue are often arranged in precise patterns of staggering complexity. For instance, the hundreds of different types of neurons in the human brain are interconnected to one another through a network of some 10^{15} synaptic connections! The coordinated functioning of many types of cells within tissues, and of multiple specialized tissues, permits the organism as a whole to move, metabolize, reproduce, and carry out other essential activities.

A key step in the evolution of multicellularity must have been the ability of cells to contact tightly and interact specifically with other cells. Various integral membrane proteins, collectively termed cell-adhesion molecules (CAMs), enable many animal cells to adhere tightly and specifically with cells of the same, or similar, type; these interactions allow populations of cells to segregate into distinct tissues. Following aggregation, cells elaborate specialized cell junctions that stabilize these interactions and promote local communication between adjacent cells. Animal cells also secrete a complex network of proteins and carbohydrates, the extracellular matrix (ECM), that creates a special environment in the spaces between cells. The matrix helps bind the cells in tissues together and is a reservoir for many hormones controlling cell growth and differentiation. The matrix also provides a lattice through which cells can move, particularly during the early stages of differentiation. Defects in these connections lead to cancer and developmental malformations.

The extracellular matrix has three major protein components: highly viscous proteoglycans, which cushion cells; insoluble collagen fibers, which provide strength and resilience; and soluble multiadhesive matrix proteins, which bind these components to receptors on the cell surface. Different combinations of these components tailor the strength of the extracellular matrix for different purposes. For example, animals contain many types of extracellular matrices, each specialized for a particular function such as strength (in a tendon), cushioning (in cartilage), or adhesion. In the case of smooth muscle cells that surround an artery, the extracellular matrix must provide strong but flexible connections.

The extracellular matrix is not just an inert framework or cage that supports or surrounds cells. The matrix also communicates directly and indirectly with the intracellular signaling pathways that direct a cell to carry out specific functions. For example, the ability of hepatocytes—the principal cells in the liver—to express liver-specific proteins depends on their association with a matrix of appropriate composition. Specific ECM components can directly activate cytosolic signal-transduction pathways by binding to cell-adhesion protein receptors in the plasma membrane.

Alternatively, by binding growth factors and other hormones, the matrix can either sequester these signals from cells or, conversely, present them to cells, thereby indirectly inducing or inhibiting signaling pathways. Morphogenesis—the later stage of embryonic development during which form is achieved by cell movements and rearrangements—also is critically dependent on ECM components, which are constantly being remodeled, degraded, and resynthesized locally. Even in adults—in areas of wounding, for example—degradation and resynthesis of ECM components occurs. How the extracellular matrix regulates cell activities is considered in other chapters on signaling and development.

In this chapter, we focus on the structure of and interactions between ECM components, cell-adhesion molecules, and cell-adhesion junctions—the main structures that permit animal cells to form organized tissues. Plant cells are surrounded by a cell wall that is thicker and more rigid than the extracellular matrix. Although the plant cell wall and the extracellular matrix serve many of the same functions, they are structurally very different. Because of these differences, we discuss the plant cell wall and its interactions with plant cells in a separate section at the end of the chapter.

Integrins

The major platelet integrin, IIb3, is required for platelet interactions with proteins in plasma and the extracellular matrices (ECMs) that are essential for platelet adhesion and aggregation during hemo stasis and arterial thrombosis. Lig and binding to IIb3 is controlled by inside-out signals that modulate receptor conformation and clustering. In turn, ligand binding triggers outside-in signals through IIb3 that, when disrupted, can cause a bleeding diathesis. In the past 5 years there has been an explosion of knowledge about the structure and function ofIIb3 and the related integrin, V3. These developments are discussed here, and current models of bidirectional IIb3 signaling are presented as frameworks for future investigations. An understan ding that IIb3 functions as a dynamic molecular scaffold for extracellular and intracellular proteins has translated into diagnostic and therapeutic insights relevant to hematology and cardiovascular medicine, and further advances can be anticipated. (Blood. 2004;104:1606-1615)

Angiogenesis

Angiogenesis refers to the formation of new blood vessels, which are required for tumor growth.

Lymphangiogenesis

The high mortality rates associated with cancer can be attributed to the metastatic spread of tumor cells from the site of their origin. Tumor cells invade either the blood or lymphatic vessels to access the general circulation and then establish themselves in other tissues. Clinicopathological data suggest that the lymphatics are an initial route for the spread of solid tumors. Detection of sentinel lymph nodes by biopsy provides significant information for staging and designing therapeutic regimens. The role of angiogenesis in facilitating the growth of solid tumors has been well established, but the presence of lymphatic vessels and the relevance of lymphangiogenesis to tumor spread are less clear. Recently, the molecular pathway that signals for lymphangiogenesis and relatively specific markers for lymphatic endothelium have been described allowing analyses of tumor lymphangiogenesis to be performed in animal models. These studies demonstrate that tumor lymphangiogenesis is a major component of the metastatic process and implicate members of the VEGF family of growth factors as key mediators of lymphangiogenesis in both normal biology and tumors.—Stacker, S. A., Baldwin, M. E., Achen, M. G. The role of tumor lymphangiogenesis in metastatic spread.

Others

Extracellular matrix (ECM) is known to provide signals controlling cell shape, migration, proliferation, differentiation, morphogenesis, and survival. Recent data shows that some of these signals are derived from biologically active cryptic sites within matrix molecules (matricryptic sites) that are revealed after structural or conformational alteration of these molecules. We propose the name, matricryptins, for enzymatic fragments of ECM containing exposed matricryptic sites. Mechanisms regulating the exposure of matricryptic sites within ECM molecules include the major mechanism of enzymatic breakdown as well as others including ECM protein multimerization, adsorption to other molecules, cell-mediated mechanical forces, and ECM denaturation. Such matrix alterations occur during or as a result of tissue injury, and thus, the appearance of matricryptic sites within an injury site may provide important new signals to regulate the repair process. Here, we review the data supporting this concept and provide insight into why the increased exposure of matricryptic sites may be an important regulatory step in tissue responses to injury.

THE IMMUNE SYSTEM

The immune system is a network of cells, tissues, and organs that work together to defend the body against attacks by "foreign" invaders. These are primarily microbes (germs) - tiny, infection-causing organisms such as bacteria, viruses, parasites, and fungi. Because the human body provides an ideal environment for many microbes, they try to break in. It is the immune system's job to keep them out or, failing that, to seek out and destroy them. When the immune system hits the wrong target or is crippled, however, it can unleash a torrent of diseases, including allergy, arthritis, or AIDS.

The immune system is amazingly complex. It can recognize and remember millions of different enemies, and it can produce secretions and cells to match up with and wipe out each one of them. The secret to its success is an elaborate and dynamic communications network. Millions and millions of cells, organized into sets and subsets, gather like clouds of bees swarming around a hive and pass information back and forth. Once immune cells receive the alarm, they undergo tactical changes and begin to produce powerful chemicals. These substances allow the cells to regulate their own growth and behaviour, enlist their fellows, and direct new recruits to trouble spots.

ORGANS OF THE IMMUNE SYSTEM

The organs of the immune system are positioned throughout the body. They are called *lymphoid organs* because they are home to *lymphocytes,* small white blood cells that are the key players in the immune system.

Bone marrow, the soft tissue in the hollow centre of bones, is the ultimate source of all blood cells, including white blood cells destined to become immune cells. The *thymus* is an organ that lies behind the breastbone; lymphocytes known as *T lymphocytes,* or just *"T cells,"* mature in the thymus.

Lymphocytes can travel throughout the body using the blood vessels. The cells can also travel through a system of lymphatic vessels that closely parallels the body's veins and arteries. Cells and fluids are exchanged between blood and lymphatic vessels, enabling the lymphatic system to monitor the body for invading microbes. The lymphatic vessels carry *lymph,* a clear fluid that bathes the body's tissues.

Small, bean-shaped lymph nodes are laced along the lymphatic vessels, with clusters in the neck, armpits, abdomen, and groin. Each lymph node contains specialized compartments where immune cells congregate, and where they can encounter antigens.

Immune cells and foreign particles enter the lymph nodes via incoming lymphatic vessels or the lymph nodes' tiny blood vessels. All lymphocytes exit lymph nodes through outgoing lymphatic vessels. Once in the bloodstream, they are transported to tissues throughout the body. They patrol everywhere for foreign antigens, then gradually drift back into the lymphatic system, to begin the cycle all over again.

The spleen is a flattened organ at the upper left of the abdomen. Like the lymph nodes, the spleen contains specialized compartments where immune cells gather and work, and serves as a meeting ground where immune defenses confront antigens.

CELLS OF THE IMMUNE SYSTEM

The immune system stockpiles a huge arsenal of cells, not only lymphocytes but also cell-devouring phagocytes and their relatives. Some immune cells take on all comers, while others are trained on highly specific targets. To work effectively, most immune cells need the cooperation of their comrades. Sometimes immune cells communicate by direct physical contact, sometimes by releasing chemical messengers.

The immune system stores just a few of each kind of the different cells needed to recognize millions of possible enemies. When an antigen appears, those few matching cells multiply into a full-scale army. After their job is done, they fade away, leaving sentries behind to watch for future attacks.

All immune cells begin as immature stem cells in the bone marrow. They respond to different cytokines and other signals to grow into specific immune cell types. Some stem cells develop into myeloid progenitor cells while others become lymphoid progenitor cells.

The myeloid progenitors develop into the cells that respond early and nonspecifically to infection. Neutrophils engulf bacteria upon contact and send out warning signals. Monocytes turn into macrophages in body tissues and gobble up foreign invaders. Granule-containing cells such as eosinophils attack parasites, while basophils release granules containing histamine and other allergy-related molecules.

Lymphoid precursors develop into the small white blood cells called lymphocytes. Lymphocytes respond later in infection. They mount a more specifically tailored attack after antigen-presenting cells such as dendritic cells (or macrophages) display their catch in the form of antigen fragments. The B cell turns into a plasma cell that produces and releases into the bloodstream thousands of specific antibodies. The T cells coordinate the entire immune response and eliminate the viruses hiding in infected cells.

Chapter 8

Biotechnology and Bioinformatics

DNA CLONING

DNA cloning is a technique to reproduce DNA fragments. It can be achieved by two different approaches:

1. Cell based, and
2. Using polymerase chain reaction (PCR). In the cell-based approach, a vector is required to carry the DNA fragment of interest into the host cell. The following figure shows a typical procedure by using plasmids as the cloning vector.

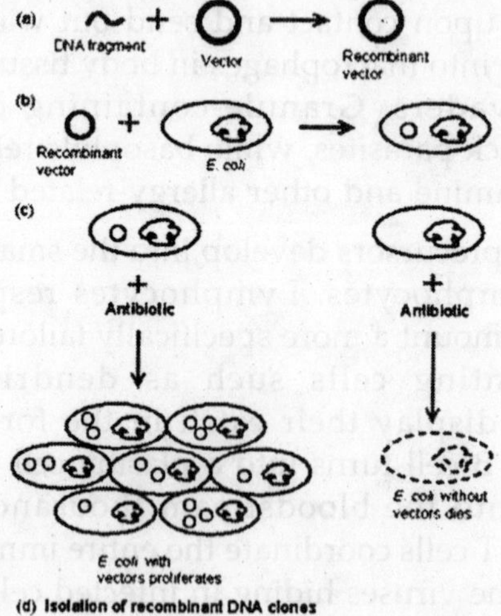

Fig. The Essential Steps in DNA Cloning using Plasmids as Vectors.

a. DNA recombination. The DNA fragment to be cloned is inserted into a vector (more information). The recombinant vector must also contain an antibiotic-resistance gene (not shown).

b. Transformation. The recombinant DNA enters into the host cell and proliferates. It is called "transformation" because the function of the host cell may be altered. Normal *E. coli* cells are difficult to take up plasmid DNA from the medium. If they are treated with $CaCl_2$, the transformation efficiency can be significantly enhanced. Even so, only one cell in about 10,000 cells may take up a plasmid DNA molecule.

c. Selective amplification. A specific antibiotic is added to kill *E. coli* without any protection. The transformed *E. coli* is protected by the antibiotic-resistance gene whose product can inactivate the specific antibiotic. In this figure, the numbers of vectors in each *E. coli* cell are not the same, because they may also reproduce independently.

d. Isolation of desired DNA clones.

DNA RECOMBINATION TECHNOLOGY

The procedure used in artificial DNA recombination is similar to the natural transpositional recombination. The major difference is that researchers can choose any appropriate enzymes to cut the DNA molecules. They are usually isolated from bacteria. The role of these enzymes in bacteria is to "restrict" the invasion of foreign DNA by cutting it into pieces. Hence, these enzymes are known as restriction enzymes.

Note: The "Recognition site" of a restriction enzyme is also called the restriction site. In this column, the first line is from 5' to 3' and the second line is from 3' to 5'. The arrow indicates the cleavage site. If the cleavage site is not at the centre, the restriction enzyme will generate sticky ends (web link) which can base-pair with other DNA fragments cleaved by the same restriction enzyme. If the cleavage site is at the centre, the restriction enzyme will generate blunt ends.

CLONING VECTORS

"Vector" is an agent that can carry a DNA fragment into a host cell. If it is used for reproducing the DNA fragment, it is called a "cloning vector". If it is used for expressing certain gene in the DNA fragment, it is called an "expression vector".

Commonly used vectors include plasmid, Lambda phage, cosmid and yeast artificial chromosome (YAC).

Plasmid

Plasmids are circular, double-stranded DNA molecules that exist in bacteria and in the nuclei of some eukaryotic cells. They can replicate independently of the host cell. The size of plasmids ranges from a few kb to near 100 kb.

A plasmid vector is made from natural plasmids by removing unnecessary segments and adding essential sequences. To clone a DNA sample, the same restriction enzyme must be used to cut both the vector and the DNA sample. Therefore, a vector usually contains a sequence (polylinker) which can recognize several restriction enzymes so that the vector can be used for cloning a variety of DNA samples.

A plasmid vector must also contain a drug-resistance gene for selective amplification. After the vector enters into a host cell, it may proliferate with the host cell. However, since the transformation efficiency of plasmids in *E. coli* is very low, most *E. coli* cells that proliferate in the medium would not contain the plasmids. Therefore, we must find a way to allow only the transformed *E. coli* to proliferate. Typically, antibiotics are used to kill *E. coli* cells which do not contain the vectors. The transformed *E. coli* cells are protected by the ampicillin-resistance gene (amp^r) which can express the enzyme, b-lactamase, to inactivate the antibiotic ampicillin.

Lambda Phage

l phages are viruses that can infect bacteria. The major advantage of the l phage vector is its high transformation efficiency, about 1000 times more efficient than the plasmid vector.

The extreme ends of the l DNA are known as COS sites, each is single stranded, 12 nucleotides long. Because their sequences are complementary to each other, one end of l DNA may base-pair with the other end of a different l DNA, forming concatemers. The two ends of a l DNA may also bind together, forming a circular DNA. In the host cell, the l DNA circularizes because ligase may seal the join of the COS sites.

In the assembly process of l virions, two proteins Nu1 and A can recognize the COS site, directing the insertion of the l DNA between them into an empty head. The filled head is then attached to the tail, forming a complete l virion. The whole process normally takes place in the host cell. However, to prepare the l virion carrying recombinant l DNA, the following in vitro assembly system is commonly used.

Proteins Nu1 and A are encoded by the genes in the l genome. If the two genes are mutated, l DNA cannot be packaged into the pre-assembled head. Because tails attach only to filled heads, the cell will accumulate separate empty heads and tails, which can then be extracted. When the extract is mixed with recombinant l DNA and proteins Nu1 and A, the complete l virion carrying recombinant l DNA will be assembled.

Cosmid

The cosmid vector is a combination of the plasmid vector and the COS site which allows the target DNA to be inserted into the l head. It has the following advantages:

- High transformation efficiency.
- The cosmid vector can carry up to 45 kb whereas plasmid and l phage vectors are limited to 25 kb.

YAC

The yeast artificial chromosome (YAC) vector is capable of carrying a large DNA fragment (up to 2 Mb), but its transformation efficiency is very low.

Essential Components of YAC Vectors

- Centromers (CEN), telomeres (TEL) and autonomous replicating sequence (ARS) for proliferation in the host cell.

- *amp*[r] for selective amplification and markers such as TRP1 and URA3 for identifying cells containing the YAC vector.
- Recognition sites of restriction enzymes (e.g., EcoRI and BamHI)

Procedure

1. The target DNA is partially digested by EcoRI and the YAC vector is cleaved by EcoRI and BamHI.
2. Ligate the cleaved vector segments with a digested DNA fragment to form an artificial chromosome.
3. Transform yeast cells to make a large number of copies.

GENOMIC AND CDNA LIBRARIES

Suppose you have known the partial sequence of a gene (e.g., from the sequence of a homologous gene) and want to determine its entire sequence, then you may use the technique described in this section.

PREPARATION OF A DNA LIBRARY

DNA library is a collection of cloned DNA fragments. There are two types of DNA library:

The genomic library contains DNA fragments representing the entire genome of an organism.

The cDNA library contains only complementary DNA molecules synthesized from mRNA molecules in a cell.

GENOMIC LIBRARY

The genomic library is normally made by l phage vectors, instead of plasmid vectors, for the following reasons:

The entire human genome is about 3 x 10^9 bp long while a plamid or l phage vector may carry up to 20 kb fragment. This would require 1.5 x 10^5 recombinant plasmids or l phages. When plating *E. coli* colonies on a 3" petri dish, the maximum number to allow isolation of individual colonies is about 200 colonies per dish. Thus, at least 700 petri dishes are required to construct a human genomic library. By contrast, as many as 5×10^4 l phage plagues can be screened on a typical petri dish. This requires only 30 petri dishes to construct a human genomic library. Another advantage of l phage vector is that its transformation efficiency is about 1000 times higher than the plasmid vector.

CDNA LIBRARY

The advantage of cDNA library is that it contains only the coding region of a genome. To prepare a cDNA library, the first step is to isolate the total mRNA from the cell type of interest. Because eukaryotic mRNAs consist of a poly-A tail, they can easily be separated. Then the enzyme reverse transcriptase is used to synthesize a DNA strand complementary to each mRNA mlecule. After the single-stranded DNA molecules are converted into double-stranded DNA molecules by DNA polymerase, they are inserted into vectors and cloned.

PROBES

A probe is a piece of DNA or RNA used to detect specific nucleic acid sequences by hybridization (binding of two nucleic acid chains by base pairing). They are radioactively labeled so that the hybridized nucleic acid can be identified by autoradiography.

The size of probes ranges from a few nucleotides to hundreds of kilobases. Long probes are usually made by cloning. Originally they may be double-stranded, but the working probes must be single-stranded. Short probes (oligonucleotide probes) can be made by chemical synthesis. They are single-stranded.

Suppose we have cloned a specific gene in yeast and want to find its homologous gene in human, then we may use the specific yeast gene as a probe to detect its homologous gene from the human genomic library. On the other hand, if we know the conserved sequence in the specific gene between yeast and human, we may use oligonucleotide probes containing only the conserved sequence. Typically, an oligonucleotide about 20 nucleotides long is sufficient to screen a library.

In some cases, we have known the partial sequence of a protein and want to detect its gene in the library. Then we may synthesize oligonucleotide probes based on the known peptide sequence. Since an amino acid may be encoded by several DNA triplets, many different oligonucleotide probes are often needed.

SCREENING

Once a particular DNA fragment is identified, it can be isolated and amplified to determine its sequence. If we know the partial sequence of a gene and want to determine its entire sequence, the probe should contain the known sequence so that the detected DNA fragment may contain the gene of interest.

GEL ELECTROPHORESIS

Gel electrophoresis is a technique for separating charged molecules with different sizes. Two kinds of gels are commonly used: agarose and polyacrylamide. Agarose gels can be applied to a wider range of sizes than polyacrylamide gels. By using standard agarose electrophoresis, nuclei acids up to 50 kb may be separated. If Pulsed Field Gel Electrophoresis is used, the upper limit can be extended to 10 Mb. Polyacrylamide gels may separate nucleic acids that differ in length by only 1 nucleotide if their length is less than 500 bp.

BLOTTING METHODS AND APPLICATIONS

Following gel electrophoresis, probes are often used to detect specific molecules from the mixture. However, probes cannot be applied directly to the gel. The problem can be solved by three types of blotting methods: Southern blotting, Northern blotting and Western blotting.

SOUTHERN BLOTTING

Southern blotting is a technique for detecting specific DNA fragments in a complex mixture. The technique was invented in mid-1970s by Edward Southern. It has been applied to detect Restriction Fragment Length Polymorphism (RFLP) and Variable Number of Tandem Repeat Polymorphism (VNTR). The latter is the basis of DNA fingerprinting.

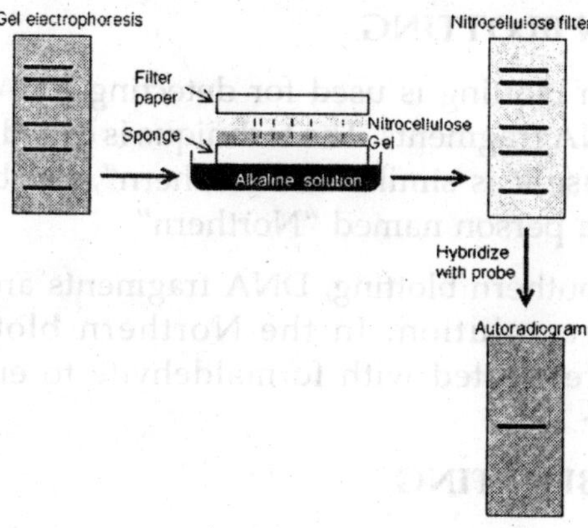

Fig. Southern Blotting.

a. The DNA to be analyzed is digested with restriction enzymes and then separated by agarose gel electrophoresis.
b. The DNA fragments in the gel are denatured with alkaline solution and transferred onto a nitrocellulose filter or nylon membrane by blotting, preserving the distribution of the DNA fragments in the gel.
c. The nitrocellulose filter is incubated with a specific probe. The location of the DNA fragment that hybridizes with the probe can be displayed by autoradiography.

RESTRICTION FRAGMENT LENGTH POLYMORPHISM (RFLP)

Polymorphism refers to the DNA sequence variation between individuals of a species. If the sequence variation occurs at the restriction sites, it could result in RFLP. The most well known example is the RFLP due to b globin gene mutation.

VARIABLE NUMBER OF TANDEM REPEAT (VNTR) POLYMORPHISM

VNTR may result from unequal crossover. It is the molecular basis of DNA fingerprinting which has many practical applications.

NORTHERN BLOTTING

Northern blotting is used for detecting RNA fragments, instead of DNA fragments. The technique is called "Northern" simply because it is similar to "Southern", not because it is invented by a person named "Northern".

In the Southern blotting, DNA fragments are denatured with alkaline solution. In the Northern blotting, RNA fragments are treated with formaldehyde to ensure linear conformation.

WESTERN BLOTTING

Western blotting is used to detect a particular protein in a mixture. The probe used is therefore not DNA or RNA, but antibodies. The technique is also called "immunoblotting".

POLYMERASE CHAIN REACTION (PCR)

PCR is a cell-free method of DNA cloning. It is much faster and more sensitive than cell-based cloning.

Materials Required

- Two primers, each about 20 bases long with sequence complementary to the sequence immediately adjacent to the DNA segment of interest.
- DNA polymerase (e.g., *Tag polymerase*) which can sustain high temperature (> 60° C)
- A large number of free deoxynucleotides (dNTPs)
- The target DNA fragment.

Procedure

1. Heat denaturation at about 95°C.
2. Primers bind to the denatured DNA by base pairing as the temperature is gradually cooled to about 60° C.
3. Extend primers with *Tag polymerase.*
4. Repeat the above process. The number of copies doubles in each cycle. Typically 20 to 30 cycles are sufficient for effective DNA amplification.

Advantages

- Much faster than using vectors.
- Only very small amount of target DNA is needed.

Disadvantages

- To synthesize primers, we need to know the sequence flanking the DNA segment of interest.
- Applies only to short DNA fragments, typically less than 5 kb.

DNA SEQUENCING

The Sanger method being used today was pioneered by Fred Sanger in the 1970s. It is also known as the Dideoxy method, because a small amount of dideoxynucleotides are mixed with normal deoxynucleotides during sequencing. The dideoxynucleotides lack both 2' and 3' hydroxyl groups, while the deoxynucleotides lack only the 2' hydroxyl group.The 3' hydroxyl group is essential for forming the phosphodiester bond that connects two nucleotides. Therefore, the dideoxynucleotide will be the terminator of a polynucleotide chain since it lacks the essential 3' hydroxyl group.

The Sanger method may sequence a DNA fragment containing up to 500 nucleotides. For large scale sequencing (such as the entire human genome), a strategy known as the shotgun sequencing is commonly used. In this approach, the DNA molecule of interest is randomly chopped into numerous small pieces. After these small pieces are sequenced, the whole sequence is assembled by their common overlaps.

SITE-DIRECTED MUTAGENESIS

Site-directed mutagenesis has been widely used in the study of protein functions. There are many approaches. An oligonucleotide-based method is illustrated in figure. This method was first developed by Michael Smith who was awarded a Nobel Prize in 1993 for this contribution. Site-directed mutagenesis can also be achieved by using PCR (more info).

1. Cloning the DNA of interest into a plasmid vector.
2. The plasmid DNA is denatured to produce single strands.
3. A synthetic oligonucleotide with desired mutation (point mutation, deletion, or insertion) is annealed to the target region. In this figure, the T to G point mutation is used as an example.
4. Extending the mutant oligonucleotide using a plasmid DNA strand as the template.
5. The heteroduplex is propagated by transformation in *E. coli*.

After propagation, in theory, about 50per cent of the produced heteroduplexes will be mutants and the other 50per cent will be the "wild type" (no mutation). In commercial mutagenesis kits, some selection and enrichment methods have been used to favour the production of mutants.

PRODUCTION OF RECOMBINANT PROTEINS

Many proteins which may be used for medical treatment or for research are normally expressed at very low concentrations. Through recombinant DNA technology, a large quantity of proteins can be produced. This involves inserting the desired protein gene into an "expression vector" which must contain a promoter so that the protein can be expressed.

ORGANISM CLONING

DNA cloning is a technique to reproduce identical DNA fragments while organism cloning (also called reproductive cloning) is to reproduce an organism with genetic materials identical to its "single" parent. During natural reproduction process, bacteria can clone themselves, but the offspring of higher organisms usually contains the mixture of genetic materials from both parents. Dolly, a sheep created by Ian Wilmut's group at the Roslin Institute in 1996, was the first cloned animal. She was created from a cell of another sheep (Dolly's single parent), rather than the combination of sperm and egg.

TECHNIQUES

Conceiving a Clone

The details section contains all the information you will ever need on cloning and biotechnology. Follow the development of cloning on the timeline, learn about the various techniques used to create clones, view animations of important processes, and read about the scientists whose accomplishments have made cloning a reality.

The recent cloning developments have created a large scale debate on the morality and ethics of duplicating humans and other creatures. Enter the reactions section to learn about common cloning misconceptions, views from both sides of the cloning debate, and what the government is doing to prevent human cloning research.

After learning about cloning and related biotechnology, step into our cloning discussions and let your views be known. Or enter our science lab and perform your own cloning experiment on the web. If you feel you are a cloning expert, test your knowledge by taking our cloning questionnaire.

Creating a Cloned Sheep Named Dolly

Who would have thought that a Scottish sheep named Dolly could cause an international uproar? But she has, and it's all because of her unorthodox beginnings. Dolly is not an ordinary sheep, produced through the mating of a ewe and a ram. She is a clone, an exact genetic replica of her donor "mother," a six-year-old female sheep. And that fact, together with another recent announcement that scientists in Oregon have cloned monkeys from embryonic cells, has triggered debates among researchers, ethicists, and politicians all over the world.

The ethical controversy over any future experiments that involve the cloning of human beings — which has NOT occurred — is likely to dominate news coverage of the subject for months to come. Before engaging in such a debate, it is important to understand what the process of cloning involves, and why Dolly is so special. It is also important to understand some of the key questions that Dolly raises.

CLONE

The noun "clone" and the verb "to clone" are not used consistently. In biology, a clone is a cell or an organism that is genetically identical to another cell or organism. Many simple organisms such as bacteria reproduce themselves by copying their DNA and splitting in half. The two bacteria that result from this form of asexual reproduction are genetically identical; they are clones of each other. In contrast, during the process of sexual reproduction, the nucleus of a sperm cell, which carries the father's DNA, fuses with the nucleus of the egg cell, which contains the mother's DNA. The resulting offspring carry genetic material from both parents and are not identical to either parent.

The verb "to clone" refers to the process of creating cloned cells or organisms. The process differs, depending on the kinds of cells used in the cloning procedure and the desired result. Usually, when scientists clone an animal, they take the nucleus of a cell — which contains chromosomes made of deoxyribonucleic acid (DNA) and proteins — and place it into an egg cell (also called an oocyte) from which the nucleus has been removed. The egg cell then divides to produce an embryo that develops into an animal, if the procedures work as planned.

In previous cloning experiments with cattle and mice, the "donor" chromosomes came from very early embryos or, in the case of cloned frogs, from young tadpoles. Very early embryos from mammals are called blastocysts, balls of immature cells that are produced after a fertilized egg divides several times, but before any specialized tissues begin to develop. Recently, two Rhesus monkeys were produced from embryonic cells in blastocysts by a method called nuclear transfer

Creation of Dolly

Dolly is different. She was generated from a specialized adult cell, not from an unspecialized embryonic cell.

To create Dolly, Ian Wilmut of the Roslin Institute in Edinburgh, Scotland, and his colleagues used a cell derived from the udder of a six-year-old sheep in the final stage of pregnancy. The researchers fused the adult udder cell with an oocyte that was ready to be fertilized, but taken from a different sheep. The scientists had previously removed the nucleus from the oocyte, and they used an electrical current to fuse it with the udder cell. The key to Dolly's success was to make the nucleus of the donor cell "quiet" so that it stopped behaving like an udder cell and could be reprogrammed to become an embryo.

The resulting embryo — which became Dolly — carried all of the chromosomes from the donor udder cell and none of the nuclear chromosomes from the host egg cell. Therefore, Dolly is an exact genetic copy — a clone — of her donor-cell "mother."

BIOINFORMATICS

Bioinformatics is the field of science in which biology, computer science, and information technology merge into a single discipline. The ultimate goal of the field is to enable the discovery of new biological insights as well as to create a global perspective from which unifying principles in biology can be discerned. There are three important sub-disciplines within bioinformatics:

- The development of new algorithms and statistics with which to assess relationships among members of large data sets
- The analysis and interpretation of various types of data including nucleotide and amino acid sequences, protein domains, and protein structures
- The development and implementation of tools that enable efficient access and management of different types of information.

DNA MICROARRAYS AND SAGE

DNA MICROARRAYS

Recently invented methods allow researchers to analyze the expression of thousands of genes simultaneously using DNA microarrays. Coupling these methods with the results from genome sequencing projects allows researchers to analyze the complete transcriptional programme of an organism during specific physiological responses or developmental processes. DNA microarrays consist of thousands of individual gene sequences bound to closely spaced regions on the surface of a glass microscope slide. Two methods have been developed for preparing DNA microarrays.

In one method, an H"1-kb portion of the coding region of each gene analyzed is individually amplified by PCR. A robotic device is used to apply each amplified DNA sample to closely spaced spots on the surface of a glass microscope slide, which then is processed by chemical and heat treatment to attach the DNA sequences to the glass surface and denature them. Typical arrays are 2 × 2 cm and contain H"6000 spots of DNA.

In an alternative method, multiple DNA oligonucleotides, usually 20 nucleotides in length, are synthesized from an initial nucleotide that is covalently bound to the surface of a glass slide. Tens of thousands of identical oligonucleotides are synthesized in a small square area on the surface of the slide. Several oligonucleotide sequences from a single gene are synthesized in neighboring regions of the slide to analyze expression of that gene. Thousands of genes can be represented on one glass slide. Because the methods for constructing these arrays of synthetic oligonucleotides were adapted from methods for manufacturing microscopic integrated circuits used in computers, these types of oligonucleotide microarrays are often called *DNA chips*.

In one of the first uses of DNA microarray analysis, gene expression was compared between yeast cells growing on glucose as the source of carbon and energy, and cells growing on ethanol. The total mRNA was separately isolated from cells growing under these two conditions using oligo-dT columns. The total mRNA from the glucose-grown cells was reverse-transcribed into cDNA using reverse transcriptase, an oligo-dT primer, deoxyribonucleoside triphosphates, and a low concentration of a nucleotide analog labeled with a green fluorescent dye. cDNA from the ethanol-grown cells was prepared similarly, but labeled with a red fluorescent dye. Equal amounts of the two cDNA preparations were then mixed and incubated under hybridization conditions with a DNA microarray containing all 6400 yeast genes, so that each gene on the array hybridized to its complementary cDNA in the preparation.

After washing away unhybridized cDNA, each DNA spot on the microarray was analyzed using a scanning laser microscope with a computer-controlled stage. The intensity of green and red fluorescence at each DNA spot was mea-sured with photomultiplier tubes and stored in computer files under the name of each gene according to its known position on the slide. The data is represented in the computer-generated image shown in figure. Spots fluorescing bright red are from genes expressed at a high level in ethanol and a low level in glucose media. Spots fluorescing bright green are from genes expressed at a high level in glucose and a low level in ethanol medium. Spots fluorescing bright yellow are from genes expressed at high levels in both types of media. Genes expressed at low to moderate levels do not yield brightly fluorescing spots in the depiction of the data in figure where all genes are compared. However, the relative level of expression of such genes in cells grown in glucose or ethanol can be determined from accurate mea-surements of the relative intensities of weak red and green fluorescence signals. Genes that are not transcribed under these growth conditions give no detectable signal.

Hybridization of fluorescently labeled cDNA preparations to DNA microarrays such as this provide a means for analyzing gene expression patterns on a gęnomic scale. One example is the analysis of changes in gene expression as yeast cells convert from growth on a high-sugar medium to growth on ethanol. This adaptation to different nutrients occurs in the natural environment when yeast growing on decaying fruit switch from rapid growth, when they metabolize sugars to ethanol by anaerobic fermentation, to slower growth, when the sugar is depleted and they metabolize the accumulated ethanol to carbon dioxide and water by aerobic respiration. This situation can be simulated in the laboratory by inoculating yeast into media high in glucose. As shown in Figure growth initially is exponential and rapid until the glucose is exhausted; then the rate of cell doubling is slowed and cells shift from fermentation to respiration. The transcriptional response of each of the 6400 yeast genes during this adaptation to decreasing glucose was analyzed by hybridizing fluorescently labeled cDNAs prepared from cells at several time points after inoculation of the culture to DNA microarrays as shown in figure.

This analysis revealed a marked, coordinated change in the global pattern of yeast gene expression. For example, expression from seventeen genes encoding mitochondrial proteins required for respiration increased with a similar time course as glucose was depleted. Analysis of the yeast genome sequence showed that each of these genes contains in their upstream transcription-control region binding sites for the trimeric transcription factor encoded by genes *HAP2,3,* and *4*. Earlier studies showed that this factor regulates the transcription of some of these genes. This genome-wide expression analysis revealed that these seventeen genes are probably chiefly regulated by the trimeric Hap transcription factor. As the rate of cell growth slowed when glucose was depleted from the media, 112 genes encoding ribosomal proteins were repressed with similar kinetics. The upstream control regions in all of these genes have binding sites for the Rap1 transcription factor, which earlier had been shown to regulate the expression of a handful of ribosomal protein genes. Seven genes exhibited strong induction only at the last time point analyzed, when glucose was completely depleted from the media. Six of these were found to have an upstream activating sequence in common, the CSRE carbon-source response element. Thus, several distinct temporal patterns of expression were recognized. In most cases, common regulatory mechanisms could be inferred for sets of genes with similar expression profiles.

The results of this study showed that as yeast adapt to decreasing glucose levels, expression of 710 genes increases by a factor of two or more, while expression of 1030 genes decreases by a factor of two or more. About 400 of the differentially expressed genes have no known function. This analysis provides the first clue as to their possible function in yeast biology.

Approximately 15,000 human genes have now been identified. Microarrays of these human genes have been prepared and used to characterize their expression in different tissues and organs, and in cancer cells compared to their normal cell counterparts. When the complete set of genes for humans and model experimental organisms such as the mouse and *Drosophila* become available, microarray technology will be used to analyze the global genome-wide gene expression patterns that occur during cellular differentiation and disease.

SAGE

Although microarray technology has huge potential, it suffers from the drawback that it is very expensive and the hardware to produce DNA chips is not readily available. An alternative approach which permits simultaneous tracking of the expression of large numbers of transcripts is the SAGE (serial analysis of gene expression) method devised by Velculescu *et al.* (1995). SAGE is a much more accessible method because it does not require any sophisticated equipment to track gene expression. It is based on two principles:

- *A short nucleotide sequence tag can uniquely identify the transcript from an individual gene provided it is from a defined position within the transcript.* For example, although the total number of human genes is expected to be of the order of 100 000, a sequence tag of only 9 bp can, in principle, distinguish 4^9 = 262 144 different transcripts.
- *Concatenation of short sequence tags allows the efficient analysis of transcripts in a serial manner.* The tags from different transcripts can be covalently linked together within a single clone and the clone can then be sequenced to identify the different tags in that clone.

To perform a SAGE analysis, poly $(A)^+$ RNA is first extracted from the source to be investigated and converted to cDNA using a biotinylated oligo (dT) primer. The resulting cDNA is cleaved with a frequent 4 base-pair cutter restriction nuclease (termed the anchoring enzyme) and liberated 32 end fragments, which will contain a biotin group, are bound to a streptavidin bead. The streptavidin-bound cDNA is split into two fractions, and the two fractions are separately ligated *en masse* to two double-stranded oligonucleotide adaptors (linkers) A and B, each designed to contain the following: an overhang which is complementary in sequence to overhanging ends generated by the anchoring enzyme; a five base recognition sequence for an enzyme which is termed the tagging enzyme and which is a type IIS restriction nuclease (one which unlike standard type II restriction nucleases cleaves at a defined distance up to 20 bp from its recognition site); sufficient additional sequence to design a specific PCR primer.

The linkered cDNA is cleaved with the chosen tagging enzyme and so the only portions of the original cDNA sequences which will remain bound to the linkers are short sequence tags. The two pools of linkers are then ligated to each other and PCR amplification is carried out using primers specific for linkers A and B. The amplified product will be a complex population of sequences each containing two adjacent sequence tags (ditags) derived from the sequences individually coupled to linkers A and B. The individual ditags can then be released by cleavage with the anchoring enzyme and the different ditags can be concatenated by simple ligation and then cloned. Sequencing of the resulting clones can then produce long read-outs of sequential ditags of fixed length.

Although a considerable fraction of human genes remain to be characterized, SAGE provides a measure of transcript quantitation since genes that are expressed strongly in a specific tissue will frequently be recovered as sequence tags. In the original experiment, Velculescu *et al.* (1995) reported recovery of 840 sequence tags from pancreatic cDNA. Of these, 498 represented 77 different transcripts and the most abundant transcripts were all produced from genes known to have pancreatic function (e.g. procarboxypeptidase A1 was represented 64 times, pancreatic trypsinogen 2 was represented 46 times, etc.).

SAGE can be expected to be useful in comparative expression studies to identify differences in gene expression between two or more cellular sources of RNA. A big advantage of the method is its speed: all genes expressed at 0.5per cent of total mRNA in one tissue but not abundant in another could be found in a single day (approximately 1000 sequence tags from each tissue).

Chapter 9

Genes and Diseases

GENE MAPPING

"Gene mapping" refers to the mapping of genes to specific locations on chromosomes. It is a critical step in the understanding of genetic diseases. There are two types of gene mapping:

Genetic Mapping: Using linkage analysis to determine the relative position between two genes on a chromosome.

Physical Mapping: Using all available techniques or information to determine the absolute position of a gene on a chromosome.

The ultimate goal of gene mapping is to clone genes, especially disease genes. Once a gene is cloned, we can determine its DNA sequence and study its protein product. For example, cystic fibrosis (CF) is the most common lethal inherited disease in the United States. As many as 1 in 2500 Americans of Northern European descent carry a gene with CF. In 1985, the gene was mapped to chromosome 7q31-q32 by linkage analysis. Four years later, it was cloned by Francis Collins and his co-workers. We now know that the disease is caused by the defect of a chloride channel (web link) - the protein product of this disease gene.

LINKAGE ANALYSIS

The genetic mapping is based on the linkage between "loci" (locations of genes). If two loci are usually inherited together, they are said to be "linked". Two loci on different chromosomes are not linked, because they are usually separated by independent assortment.

A locus (singular of loci) may have different sequences, referred to as alleles. Consider two loci A and B, each having two alleles (one from mother, another from father). A_1 and A_2 are the two alleles of locus A; B_1 and B_2 are the two alleles of locus B. Initially, A_1 and B_1 are located on the same chromosome. A_2 and B_2 are located on a different chromosome.

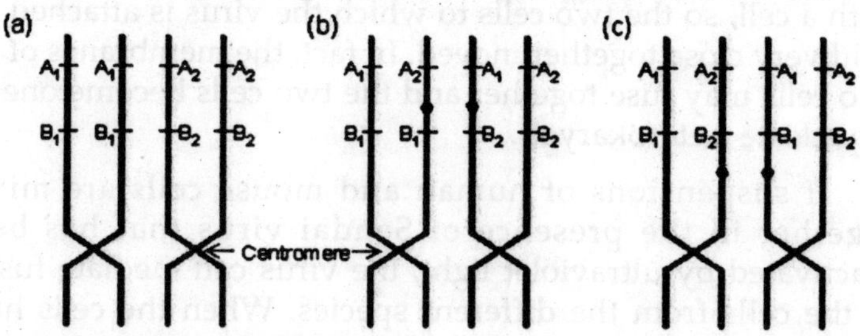

Fig. Illustration of Recombination Between Two Loci A and B.

a. Two pairs of sister chromatids align during meiosis. A_1 and B_1 are located on the same chromosome. A_2 and B_2 are located on a different chromosome.

b. DNA crossover leads to recombination if the chiasma is located between the two loci.

c. DNA crossover does not lead to recombination if the chiasma is not located between the two loci.

The DNA crossover may cause recombination of loci A and B. Namely, A_1 and B_2 (or A_2 and B_1) are located on the same chromosome. The recombination frequency depends on the distance between the two loci and the position of crossover (the chiasma). The closer they are, the less likely the recombination will occur, because recombination occurs only when the chiasma is located between the two loci.

To apply this basic principle to map a disease gene, we need to analyze the pedigree and estimate recombination frequency.

PHYSICAL MAPPING TECHNIQUES

The technique of somatic cell hybridization is extensively used in human genome mapping, but it can in principle be used in many different animal systems. The procedure uses cells growing in culture. A virus called the *Sendai virus* has a useful property that makes the mapping technique possible. Each Sendai virus has several points of attachment, so it can simultaneously attach to two different cells if they happen to be close together. However, a virus is very small in comparison with a cell, so the two cells to which the virus is attached are held very close together indeed. In fact, the membranes of the two cells may fuse together and the two cells become one—a binucleate heterokaryon.

If suspensions of human and mouse cells are mixed together in the presence of Sendai virus that has been inactivated by ultraviolet light, the virus can mediate fusion of the cells from the different species. When the cells have fused, the nuclei subsequently fuse to form a uni-nucleate cell line composed of both human and mouse chromosome sets. Because the mouse and human chromosomes are recognizably different in number and shape, the two sets in the hybrid cells can be readily distinguished. However, in the course of subsequent cell divisions, for unknown reasons the human chromosomes are gradually eliminated from the hybrid at random. Perhaps this process is analogous to haploidization in the fungus *Aspergillus*.

The loss of human chromosomes can be arrested in the following way to encourage the formation of a stable partial hybrid. The cells used are mutant for some biochemical function; so, if the cells are to grow, the missing function must be supplied by the other genome. This selective technique results in the maintenance of hybrid cells that have a complete set of mouse chromosomes and a small number of human chromosomes, which vary in number and type from hybrid to hybrid but which always include the human chromosome carrying the wild-type allele defective in the mouse genome.

Let's look at the specific genes that make the selective system work. In cells, DNA can be made either de novo ("from scratch") or through a salvage pathway that uses molecular skeletons already available. The selective technique involves the application of a chemical, aminopterin, that blocks the de novo synthetic pathway, confining DNA synthesis to the salvage pathway. Two essential salvage enzymes, thymidine kinase (TK) and hypoxanthine-guanine phosphoribosyl transferase (HGPRT), are relevant to the system, as shown in the following two reactions:

$$\textbf{thymine} \xrightarrow{\text{TK}} \textbf{thymidilic acid (a DNA precursor)}$$

$$\textbf{hypoxanthine} \xrightarrow{\text{HGPRT}} \textbf{inosinic acid (a DNA precursor)}$$

The mouse cell line to be fused is genetically unable to make TK because it is homozygous for the allele tk'', whereas the human cell line is genetically unable to make HGPRT because it is homozygous at another locus for the allele $hgprt''$. So the genotypes of the two fusing cell lines are:

Mouse: tk^-/tk^-; $hgprt^+/hgprt^+$

Human: tk^+/tk^+; $hgprt^-/hgprt^-$

Because each is deficient for one enzyme, neither the mouse nor the human cells are able to make DNA individually. In the hybrid cells, however, the tk^+ allele complements the $hgprt^+$ allele, so the cells can make both enzymes. Therefore, DNA is synthesized and the cells can proliferate. Most human chromosomes are eliminated from the hybrid cell cultures because their loss has no effect on the cultures' ability to grow. But, to continue to grow in medium containing hypoxanthine, aminopterin, and thymidine *(HAT medium)*, a hybrid culture must retain at least one of the human chromosomes that carries the tk^+ allele.

Luckily, the progressive elimination of the human chromosomes from the fused cell lines can be followed under the microscope because mouse chromosomes can easily be distinguished from human chromosomes. Moreover, chromosome stains such as quinacrine and Giemsa reveal a pattern of banding within the chromosomes. The size and the position of these bands vary from chromosome to chromosome, but the banding patterns are highly specific and invariant for each chromosome. Thus, it is easy to identify the human chromosomes that are present in any hybrid cell. Different hybrid cells are grown separately into lines; eventually a bank of lines is produced that contains, in total, all the human chromosomes.

With a complete bank of chromosomes, we can begin to assign genes or markers to chromosomes. If the human chromosome set is homozygous for a human molecular marker—such as an allele that controls a cell-surface antigen, drug resistance, a nutritional requirement, a specific protein, or a DNA marker—then the presence or absence of this genetic marker in each line of hybrid cells can be correlated with the presence or absence of certain human chromosomes in each line. Data of this sort are presented in Table in which "+" means presence and """ means absence of the genetic marker. We can see that, in the different hybrid cell lines, genetic markers 1 and 3 are always present or absent together. We can conclude, then, that they are linked. Furthermore, the presence or absence of genes 1 and 3 is correlated with the presence or absence of chromosome 2, so we can assume that these genes are located on chromosome 2. By the same reasoning, gene 2 must be on chromosome 1, but the location of gene 4 cannot be assigned. Large numbers of human genes have now been localized to specific chromosomes in this way.

The linear length of DNA in an average sized human chromosome is about 5 cm, but it is compacted by various hierarchies of folding in metaphase chromosomes to a few microns. As a result, standard *in situ* hybridization against the highly condensed chromatin of metaphase and prometaphase chromosomes does not have a high mapping resolution, with typical upper limits of about one to several megabases of DNA. To obtain higher resolution, DNA probes are hybridized to the naturally extended interphase chromosomes or to artificially stretched chromatin or DNA fibers prepared by a variety of different methods.

In situ Hybridization using Conventional Interphase Chromosomes

The chromosomes of interphase nuclei are much more extended than metaphase or prometaphase chromosomes. As a result, FISH analyses of the chromosomes of interphase nuclei can permit a high mapping resolution, and can help to determine the physical mapping order of some syntenic DNA clones. Because the extended chromosomes may loop back at certain regions, an accurate linear order for syntenic DNA clones requires a statistical analysis on the mapping results from many individual interphase nuclei. Even so, the method is best suited to ordering sequences which are separated by intervals within the 50–500 kb range.

FISH Mapping on Artificially Extended Chromatin or DNA Fibers

High resolution FISH can also be conducted using procedures which cause the DNA of chromosomes on a microscope slide to be extended prior to hybridization (extended chromatin fiber FISH. One such method, DIRVISH (direct visual hybridization), involves lysing cells with detergent at one end of a glass slide, tipping the slide, and allowing the DNA in solution to stream down the slide. Such preparations permit extremely high mapping resolutions: from over 700 kb to under 5 kb. More recently, the principle of artificially stretching chromatin fibers has been extended to protein-free DNA: the target DNA is prepared from cells embedded in pulsed field gel electrophoresis blocks. This DNA fiber FISH method used unfixed linearized DNA fibers on a microscope slide and has a resolution of from 500 kb to a few kilobases.

CLONING DISEASE GENES

MAJOR STRATEGIES

Functional cloning: Using information about the function of a known protein that could be involved in a genetic disease. This approach has very limited application.

Candidate gene approach: Using information about the gene's possible function, homology, and expression pattern, but without any knowledge of its chromosomal location.

Positional cloning: Using only information about the gene's approximate chromosomal location obtained from gene mapping.

Positional candidate approach: Using information from map position and the gene's possible function, homology, and expression pattern. This approach has been quite successful and will dominate other strategies.

Between 1986 and 1995, more than 50 human disease genes were discovered by pure positional cloning. A labour intensive and time consuming method called chromosome walking is often involved in this approach. It took ten years for a large group of researchers to clone the Huntington disease gene. Cloning of the cystic fibrosis gene took four years, which was facilitated by a special technique called chromosome jumping. In the future, as the entire human genome sequence is known, there is no need to use chromosome walking (or jumping) to identify disease genes.

MUTATION DETECTION METHODS

To confirm a candidate gene, it is often necessary to screen mutations from patient's genes. In addition to direct DNA sequencing, the following methods have also been used.

SSCP (Single Strand Conformational Polymorphism Analysis)

DGGE (Denaturing Gradient Gel Electrophoresis)

DHPLC (Denaturing High-Performance Liquid Chromatography

CCM (Chemical Cleavage of Mismatches)

EMC (Enzyme Mismatch Cleavage)

Heteroduplex analysis

DNA microarrays

GENE THERAPY

Many diseases are due to dysfunction of mutated genes. In traditional gene therapy, additional copies of therapeutic genes are delivered into transfected cells. This approach has several drawbacks (more info). Recently, researchers are focusing on a new approach: direct repair of mutated genes.

The main strategy of gene therapy has traditionally been focused on gene augmentation. This approach typically involves the introduction of an expression system designed to express a specific protein in the transfected cell. Both the basic and clinical sciences have generated enough information to suggest that gene therapy would eventually alter the fundamental practice of modern medicine. However, despite progress in the field, widespread clinical applications and success have not been achieved. The myriad deficiencies associated with gene augmentation have resulted in the development of alternative approaches to treat inherited and acquired genetic disorders. One, derived primarily from the pioneering work of homologous recombination, is gene repair. Simply stated, the process involves targeting the mutation in situ for gene correction and a return to normal gene function.

Site-specific genetic repair has many advantages over augmentation although it too is associated with significant limitations. This review outlines the advantages and disadvantages of gene correction. In particular, we discuss technologies based on chimeric RNA/DNA oligonucleotides, single-stranded and triplex-forming oligonucleotides, and small fragment homologous replacement. While each of these approaches is different, they all share a number of common characteristics, including the need for efficient delivery of nucleic acids to the nucleus. In addition, we review the potential application of a novel and exciting nonviral gene augmentation strategy—the Sleeping Beauty transposon system.

CANCER

Cancer, or malignant tumor, differs from benign tumors in two important aspects:

1. Cancer cells can divide and proliferate without control.
2. Cancer cells can invade neighboring or distant tissues, forming a secondary tumor. The original tumor is called primary tumor. The spread of cancer is known as metastasis.

Most malignant tumors originate from epithelial cells that line the surface of an organ or the body. This type of tumors are called carcinoma. Another type, called sarcoma, originates from connective tissues such as bone, muscle and blood vessels.

ABNORMAL CELL DIVISION [

Normally, cell division is triggered by growth stimulatory signals that induce the expression of cyclin D1 and other proteins involved in the progression of cell cycle from G_1 to S phase. For example, binding of epidermal growth factor (EGF) to its receptor induces the expression of c-Fos via the Ras-Raf-MEK-MAPK pathway (illustration). c-Fos is a component of the transcription factor AP-1 that regulates the expression of cyclin D1 and many proteins involved in DNA synthesis. Abnormal cell division may result if any gene involved in this process has been mutated. Important examples:

p53

p53 plays a central role at the cell cycle checkpoint to ensure replication of correct DNA. Defective p53 would allow daughter cells to contain incorrect DNA.

pRB

pRB regulates the activity of E2F, which controls the expression of many proteins involved in DNA synthesis.

Ras

Ras is involved in the signaling pathway for cell growth. Its active state is normally inactivated by rapid GTP hydrolysis. However, a mutated form cause Ras to remain in the active state for a much longer period.

EGFR

EGFR, also called ErbB, is the receptor for epidermal growth factor and its related peptide such as TGFa. Its mutated form, EGFRvIII, is activated even in the absence of ligand binding.

Mechanisms of metastasis

Epithelial cells are separated from other tissues by basal lamina (basement membrane) which is a kind of extracellular matrix (ECM). They are normally well connected. Epithelial cells are attached to each other by the molecular complex of E-cadherin and catenin, while integrin is responsible for the binding between epithelial cell and ECM. Metastasis begins with the loss of these adhesion molecules and the activation of proteolytic enzymes such as plasmin and matrix metalloproteases (MMPs) that can degrade ECM, These enable transformed cells to penetrate ECM, enter blood or lymphatic vessels, and circulate to a distant location.

BREAST CANCER

Estrogen plays a critical role in the development of breast cancer, because it can stimulate the growth of breast tissues. If the breast contains abnormal cells, they may also be stimulated by estrogen to proliferate and become a tumor.

Estrogen and Its Receptors

Two receptors for estrogen have been identified: ERα and ERβ. We shall focus on ERα since its biological functions are relatively well understood. ERα is a transcription activator which can bind to the Estrogen Response Element of the target gene and then activate the gene transcription. In the absence of estrogen, ERα is inactivated by the binding of a heat shock protein (HSP90) within the nucleus. Upon estrogen binding, the receptor undergoes conformational change so that HSP90 is released, enabling ERα to activate gene transcription. ERα may also become a transcription repressor upon binding of certain antagonists such as tamoxifen - the most widely used drug for breast cancer.

More recently, it was found that ERα can also be activated in the absence of estrogen. The structural change required for its activation can also be induced by phosphorylation of some of its residues. Since protein kinases (which catalyze phosphorylation) are activated during growth factor signaling, there is substantial "cross-talk" between growth factor and estrogen receptor signaling. The existence of estrogen-independent pathways suggests that drugs targeting estrogen alone (e.g., Aromatase inhibitors) may not be able to stop tumor growth completely. Its mechanism also provides explanation for tamoxifen resistance.

As mentioned above, a major function of estrogen is to stimulate cell growth and proliferation. This can be achieved by expressing growth factors via the function of estrogen receptor (ER) as a transcription activator. The process of gene expression, however, is slow. Recently, it has been demonstrated that estrogen can also stimulate cell proliferation without expressing any molecules. This pathway (known as nongenomic pathway) does not involve ER, but requires a G-protein-coupled receptor called GPR30. Experiments suggest that estrogen may bind to GPR30, activating metalloproteinases to cleave heparan-bound EGF (epidermal growth factor). The released EGF can then stimulate cell proliferation by acting on its receptor EGFR. In this pathway, a functional growth factor is rapidly generated by enzymatic cleavage without invoking slow gene expression. The existence of ER-independent pathway suggests that drugs targeting ER only (e.g., ER downregulators) may not be sufficient to inhibit tumor growth.

Prominent Genes in Breast Cancer

BRCA1

Mutations of BRCA1 are found in approximately 8per cent of breast cancer cells. However, if a woman is inherited with BRCA1 mutations, the possibility to develop breast cancer by age 50 is as high as 60per cent. BRCA1 is a transcription regulator. It can repress the transcriptional activation function of ERa Mutated BRCA1 would allow ERa to activate transcription of target genes, thereby promoting proliferation of abnormal cells. BRCA1 also participates in DNA repair (review). Related genes: BRCA2, BRCA3.

BRCC: A holoenzyme complex containing BRCA1, BARD1, BRCC45 and BRCC36. This complex is involved in DNA repair.

EMSY: A newly discovered protein that provides a link to BRCA2 for sporadic breast and ovarian cancer.

HER2

HER2 is also called ErbB2 or Neu, with structure and function similar to EGF receptor. This gene is overexpressed in 20-25per cent of breast cancer.

CCND1

This gene encodes the protein cyclin D1, involved in cell cycle progression. It is overexpressed in approximately 15per cent of breast cancer.

p53

Its frequency of mutation in breast cancer is 20-25per cent, well below its average of 50per cent in all types of cancer.

Ras

Its frequency of mutation in breast cancer is less than 10per cent, also below its average of 25per cent in all types of cancer.

Mechanisms of Drug Actions

Tamoxifen

Tamoxifen binds to ER and modulates its transcription function. Its usefulness is limited by the development of drug resistance: about half of patients with advanced breast cancer immediately fail to respond to tamoxifen, others eventually develop the drug resistance. Possible mechanisms:

1. As discussed above, there is an ER-independent pathway to stimulate cell proliferation while tamoxifen targets ER only.

2. Although tamoxifen was designed to inhibit ER, it may act as a partial agonist in the uterus, increasing the risk of endometrial cancer. Whether the ER is a transcription activator or repressor depends on the molecules it recruits to the promoter region of the target gene (review). This in turn depends on the structure and phosphorylation status of ER. It seems that the high activities of protein kinase (e.g., MAPK) in breast tumor may activate ER in the absence of any ligand, or even convert the tamoxifen-bound ER from a transcription repressor into an activator.
3. Overexpression of BCAR1 (breast cancer anti-estrogen resistance 1) which encodes $p130^{Cas}$, a protein involved in the regulation of signaling via Ras family.

ER Downregulators

These drugs destroy estrogen receptors (web link). Thus, they will not have drug resistance due to the mechanism 2.

SERMs

SERMs stands for "selective estrogen-receptor modulators". Examples: tamoxifen, raloxifene and arzoxifene.

Herceptin

This drug is a monoclonal antibody that binds to HER2, and inactivates its function. The drug is used primarily for patients who overexpress HER2.

Aromatase Inhibitors

These drugs bind to the body's aromastase enzyme, which is responsible for producing estrogen. Example: Femara.

Chemotherapy Drugs

These drugs attack dividing cells by altering DNA structure or interfering with microtubules. Examples: Adriamycin, Ellence, Cytoxan, and Taxol.

CERVICAL CANCER

Cervical cancer is mainly caused by human papilloma viruses (HPV), which are a group of more than 100 viruses.

Type 16 and type 18 are the high-risk HPVs for cervical cancer. About 70per cent of cervical cancer patients are infected with HPV 16 or 18.

The genomes of HPV 16 and HPV 18 encode two proteins, E6 and E7, which can inactivate two crucial proteins involved in cell cycle control: p53 and pRB. The E7 protein can bind directly with pRB and inactivate its function. The E6 protein does not associate with p53 directly. Rather, it forms a complex with a cellular protein called E6-AP. The complex can induce degradation of p53 through ubiquitination.

The E6 protein can also induce telomerase activity. High level of telomerase activity is a characteristic of cancer cells since telomerase is required for a large number of cell divisions.

Other genes involved in cervical cancer: KAI1, PTEN, and ST3.

COLORECTAL CANCER

Two pathways for the progression of colorectal cancer have been identified: chromosomal instability (CIN) and microsatellite instability (MIN or MSI). About 85per cent of colorectal cancers are due to CIN and the remaining 15per cent arise from MIN.

In the CIN pathway, the morphological change at each stage is often associated with alteration of specific genes. Initially, a polyp forms on the colon wall, which is associated with loss of the APC tumor suppressor gene. Then, the progression from the polyp into a class II adenoma is accompanied by the activation of K-Ras. Subsequently, the loss of the DCC tumor suppressor gene converts the class II adenoma into class III adenoma. Finally, a carcinoma is developed by the loss of p53 (illustration from Lodish et al.).

The MIN pathway is characterized by defects in the DNA repair genes such as *MLH1, MSH2, and MSH6* .

LEUKEMIA

Leukemia may be divided into four types:

- Acute lymphocytic leukemia (ALL): Most common in children.

- Acute myeloid leukemia (AML): Occurs in both adults and children.
- Chronic lymphocytic leukemia (CLL): Most common in adults over the age of 55.
- Chronic myeloid leukemia (CML): Occurs mainly in adults.

Fusion Genes in leukemia

Leukemia (except CLL) often arises from chromosome translocation, resulting in fusion between two genes, thereby altering the functions of gene products. The following is a list of fusion genes. Further detail can be found on this site.

ALL: AML1-ETV6, ETV6-ABL, ETV6-CDX2, ETV6-CHIC2, ETV6-JAK2, IGH-BCL2, IGH-IL3, MLL-AF4, MLL-GAS7, MLL-MLLT1, MLL-MLLT6, MLL-MSF, MYC-IGH/K/L, MYC-TRA, LMO2-TRA/B/D, TAL1-TRA/B/D, TAL1-TCTA, TCF3-HLF, TCF3-PBX1, TCL1-TRA, TCL3-TRD

AML: AML1-ETO, AML1-EVI1, ETV6-ABL, ETV6-CDX2, ETV6-CHIC2, ETV6-MN1, ETV6-EVI1, FGFR1-FOP, FUS-ERG, MLL-ABI1, MLL-AF1Q, MLL-AF4, MLL-AF6, MLL-AF9, MLL-AF10, MLL-AFX1, MLL-CBP, MLL-ELL, MLL-FBP17, MLL-GAS7, MLL-MLLT1, MLL-MSF, MLL-PNUTL1, MLL-RARA, MLL-SEPT6, MOZ-CBP, MYH11-CBFB, NPM1-MLF1, NUP98-HOXA9, NUP214-DEK, RARA-NUMA, RARA-PML, RARA-ZNF145, RBM15-MAL, PDGFRB-TRIP11, TCF3-PBX1, TCL1-TRA

CML: BCR-ABL

Alternate symbols:

ABI1 = SSH3BP1

AF4 = MLLT2

AF6 = MLLT4

AF9 = MLLT3

AF10 = MLLT10

AF17 = MLLT6

AFX1 = MLLT7

AML1 = RUNX1

ETO = MTG8 = CBFA2T1

ETV6 = TEL

MOZ = ZNF220

NUP214 = CAN

TCF3 = E2A

TCL3 = HOX11 = TLX1

Prominent Genes in CLL

CLL is the most common leukemia in adults. It is frequently associated with chromosomal deletions on 13q (50per cent), 11q (20per cent), and 17p (15per cent). The major genes in 11q and 17p are ATM and p53, respectively. However, the most important gene in 13q was not known. Recently, researchers have identified in this region a new tumor suppressor gene which could be the long-sought gene.

LIVER CANCER

Hepatocellular carcinoma (HCC) is the most common type of liver cancer, mainly due to the chronic infection of hepatitis B virus (HBV). Other risk factors include exposure to aflatoxin B_1 (AFB_1), infection of hepatitis C virus (HCV), and alcoholic cirrhosis. Recently, it has been demonstrated that the gene Foxm1 is essential for the development of HCC and is negatively regulated by the p19(ARF) tumor suppressor.

HBxAg

The genome of HBV encodes an X antigen (HBxAg) which plays a critical role in hepatocarcinogenesis. HBxAg can inactivate several tumor suppressors such as p53, p21, and pRB.

Aflatoxin B_1

Aflatoxin is produced by fungi on peanuts, corn, and other nuts and grains. Aflatoxin B_1 may interact with DNA, resulting in a G to T mutation in p53 (review).

LUNG CANCER

Lung cancer can be divided into two types: small cell lung cancer and nonsmall cell lung cancer, based on how the cells look under a microscope. Small cell lung cancer grows and spreads more quickly than nonsmall cell lung cancer. Fortunately, the former is less common.

Carcinogens

The major cause of lung cancer is exposure to carcinogens which alter the genes involved in cell division. Tobacco is the most common carcinogen, followed by asbestos.

Prominent Genes in Lung Cancer

p53, p16, MYCL1 and FHIT.

PROSTATE CANCER

Prostate cancer often begins with the loss of NKX3.1 gene, resulting in the morphological change from normal epithelium to prostatic intraepithelial neoplasia (PIN). Further loss of PTEN and Rb genes converts the PIN lesion into invasive carcinoma. The final metastasis is accelerated by the loss of p53 (illustration from Genes and Development).

Prominent Genes in Prostate Cancer

NKX3A(NKX3.1), PTEN, RB1, p53, p27, p21, PCA3, KAI1.

SKIN CANCER

There are three forms of skin cancer: basal cell carcinoma, melanoma and squamous cell carcinoma. Basal cell carcinoma is the most common of all cancers, but melanoma is the most serious form of skin cancer.

Prominent Genes in Skin Cancer

Basal cell carcinoma: PTCH, p53,

Melanoma: CDKN2A(p16), MC1R, MITF.

Squamous cell carcinoma: H-Ras, p53.

ALZHEIMER'S DISEASE

Alzheimer's disease (AD) is characterized by amyloid plaques and neurofibrillary tangles in the brain. Several molecules are involved in the disease:

- Beta amyloid (Ab)
- Presenilin
- Cholesterol
- Apolipoprotein E (ApoE)
- Tau protein

Their causal relationships have been largely unveiled. It turns out that beta amyloid plays the central role in the development of Alzheimer's disease (more info).

Beta amyloid is a protein fragment generated from a larger protein called amyloid precursor protein (APP) (more info). These fragments may clump together and mix with other molecules, neurons, and non-nerve cells to form the amyloid plaque. The length of an Ab peptide ranges from 39 to 43 amino acids. It has been found that the longer one is more likely to form plaques (thereby more toxic) than the shorter one.

ASSOCIATED GENES

Four genes have been conclusively shown to affect the development of Alzheimer's disease: the APP gene on chromosome 21, the PS1 gene on chromosome 14, the PS2 gene on chromosome 1, and the apoE gene on chromosome 19. Mutations of APP, PS1 and PS2 genes are linked to the rare early-onset form of familial AD. Many people with Down's syndrome also develop AD-like dementia by the age of 40. They have three copies of chromosome 21 (where the APP gene is located), instead of two for normal people.

APP

The APP gene encodes the precursor protein to beta amyloid. One more copy of the APP gene in people with Down's syndrome should make more precursor proteins, thereby producing more Ab peptides. Mutations of the APP gene in the early-onset AD either increase the total level of Ab or favour the production of longer (more toxic) Ab. Both cases increase the chance of forming amyloid plaques.

PS1 and PS2

PS1 and PS2 genes encode presenilin 1 and presenilin 2, respectively. They are the components of the g-secretase, which cleaves amyloid precursor protein (APP) to generate Ab. Their mutations favour the production of longer (more toxic) Ab.

ApoE

Apolipoprotein E (apoE) is a protein with 299 amino acids. Its gene is polymorphic, with three alleles, encoding three protein isoforms: apoE2, apoE3, and apoE4. Their frequencies in populations are 5–10per cent, 60–70per cent, and 15–20per cent, respectively. People with apoE4 are more likely to develop Alzheimer's disease. Experimentally, it has been found that apoE can bind to Ab, thereby inhibiting its clearance and stimulating its deposition. The apoE4 binds to Ab more rapidly than apoE3.

OBESITY AND DIABETES

OBESITY

Obesity is characterized by excessive growth of adipocytes (fat cells), which are the primary site for energy storage in the form of triglycerides (fats). A triglyceride molecule consists of three fatty acid chains and a glycerol. When the body needs energy (e.g., during exercise), some hormones can activate a hormone-sensitive lipase that is present in large quantity at the fat cell membrane. The activated lipase then causes the breakdown of triglycerides, releasing free fatty acids for the production of ATP.

Many proteins are involved in the pathogenesis of obesity. Among them, leptin plays the central role. Leptin is a 146 amino acid protein, encoded by the obesity (*ob*) gene. It is expressed mainly in adipocytes: the more fats in the cell, the more leptin molecules are produced. Leptin can counteract the buildup of fats by at least two different mechanisms:

1. Regulating enzymes involved in the synthesis of fatty acids and

2. Inhibiting appetite.

The appetite is controlled by two types of neurons in the hypothalamus of the brain: NPY/AgRP neurons and POMC/

CART neurons. Activation of the NPY/AgRP neurons releases NPY (neuropeptide Y) and AgRP (agouti-related protein), which stimulate appetite. Activation of the POMC/CART neurons releases a-MSH (a-melanocyte-stimulating hormone) and CART (cocaine and amphetamine-regulated transcript), which inhibits appetite. POMC (Pro-opiomelanocortin) is the precursor of a-MSH.

Leptin may circulate through the bloodstream to these neurons and act on its receptors, which relay the signal via the JAK/STAT pathway. In the NPY/AgRP neurons, leptin suppresses the expression of NPY and AgRP, while in the POMC/CART neurons, it enhances the expression of POMC and CART. Therefore, in both sets of neurons, leptin acts to reduce food intake, making the body leaner.

When the normal function of leptin was discovered, researchers thought that leptin might be used to treat obesity. Unfortunately, this is true only for very rare cases caused by defects in leptin or its production. For most obese people, leptin and its production are normal. In fact, since the production of leptin increases with increasing fats, most obese people have high level of leptin, but it does not induce the expected responses. This phenomenon is called leptin resistance, which could arise from certain defects in the leptin signaling cascade. Defective POMC has been found in a small number of obese people, but defects in leptin receptor, NPY or NPY receptor are rarely observed.

Recently, SOCS-3 (suppressor of cytokine signaling 3) has been demonstrated to play a key role in leptin resistance. The leptin receptor belongs to the cytokine receptor family, which signals through the JAK/STAT pathway. SOCS proteins (SOCS-1 to 7) are the negative regulators of this signaling pathway. After cytokine stimulation, they are produced to avoid excessive stimulation.When leptin is injected into normal animals, the production of SOCS-3 increases. The signaling cascade is blocked by the binding of SOCS-3 to the leptin receptor. It has also been shown that older rats have less leptin sensitivity and higher SOCS-3 expression than younger rats.

Protein tyrosine phosphatase 1B (PTP1B) is another negative regulator for the leptin signaling. Its inhibitor could be an effective drug for treating obesity by increasing leptin sensitivity. As discussed below, the type 2 diabetes is related to insulin resistance and leptin resistance. PTP1B acts as a negative regulator for both leptin and insulin signaling. Therefore, its inhibitor could be used for the two closely related diseases.

MONOGENETIC DISEASES

Disease	Gene	Protein Product
Achondroplasia	FGFR3	Fibroblast growth factor receptor 3
Adrenoleukodystrophy	ABCD1	ATP-binding cassette (ABC) transporters
Alpha-1-antitrypsin deficiency	SERPINA1	Serine protease inhibitor
Alport syndrome	COL4A5	Collagen, type IV, alpha 5
Amyotrophic lateral sclerosis	SOD1	Superoxide dismutase 1
Angelman syndrome	UBE3A	Ubiquitin protein ligase E3A
Ataxia telangiectasia	ATM	Ataxia telangiectasia mutated
Autoimmune polyglandular syndrome	AIRE	Autoimmune regulator
Burkitt lymphoma	MYC	v-myc myelocytomatosis viral oncogene homolog
Congenital adrenal hyperplasia	CYP21	cytochrome P450, family 21
Cystic fibrosis	CFTR	Cystic fibrosis transmembrane conductance regulator
Diastrophic dysplasia	SLC26A2	Sulfate transporter
Duchenne muscular dystrophy	DMD	Dystrophin

Ellis-van Creveld syndrome	EVC	EVC Protein
Familial Mediterranean fever	MEFV	Mediterranean fever protein
Fragile X syndrome	FMR1	Fragile X mental retardation 1
Friedreich's ataxia	FRDA	Frataxin
Gaucher disease	GBA	Glucosidase
Glucose galactose malabsorption	SGLT1	Sodium/glucose cotransporter
Gyrate atrophy	OAT	Ornithine aminotransferase
Hemophilia A	F8	Coagulation factor VIII
Hereditary hemochromatosis	HFE	Hemochromatosis protein
Huntington disease	HD	Huntingtin
Immunodeficiency with hyper-IgM	TNFSF5	Tumor necrosis factor member 5
Lesch-Nyhan syndrome	HPRT1	Hypoxanthine phosphoribosyltransferase 1
Maple syrup urine disease	BCKDHA	Branched chain keto acid dehydrogenase
Marfan syndrome	FBN1	Fibrillin
Menkes syndrome	ATP7A	ATPase, Cu++ transporting
Myotonic dystrophy	DMPK	Myotonic dystrophy protein kinase
Neurofibromatosis	NF1, NF2	Neurofibromin
Paroxysmal nocturnal hemoglobinuria	PIGA	Phosphatidylinositol glycan
Pendred syndrome	PDS	Pendrin
Phenylketonuria	PAH	Phenylalanine hydroxylase
Refsum disease	PHYH	Phytanoyl-CoA hydroxylase
Retinoblastoma	RB	Retinoblastoma 1
Rett syndrome	MECP2	Methyl CpG binding protein

SCID-ADA	ADA	Adenosine deaminase
SCID-X-linked	IL2RG	Interleukin-2 receptor, gamma
Sickle cell anemia	HBB	Hemoglobin, beta
Spinal muscular atrophy	SMN1	Survival of motor neuron 1
Spinocerebellar ataxia	SCA1	Ataxin 1
Tangier disease	ABCA1	ATP-binding cassette A1
Tay-Sachs disease	HEXA	Hexosaminidase
Thalassemia	HBA 1and2	Hemoglobin alpha 1 and 2
Von Hippel-Lindau syndrome	VHL	Elongin binding protein
Werner syndrome	WRN	Werner syndrome protein
Wilson's disease	ATP7B	ATPase, Cu++ transporting
Zellweger syndrome	PXR1	Peroxisome receptor 1

OTHER DISEASES

ALLERGY

There are four types of hypersensitivity diseases. Allergy is the type I characterized by high level of IgE antibodies arising from hypersensitive reaction to allergens such as pollen or food. IgE antibodies are produced from antibody-secreting plasma cells that are differentiated from B cells upon activation by allergens. The produced antibodies may bind to the Fc receptors on mast cells, which will trigger allergic reactions when the allergens interact with the bound antibodies.

The activation of B cells for the production of IgE requires two separate signals: the cytokines IL-4 or IL-13 interacting with receptors on the B-cell surface, and interaction between CD40 ligand on the helper T cell surface and CD40 on the B-cell surface. There are two types of helper T cells: T_H1 and T_H2. T_H2 is a better initiator for antibody response because it can produce IL-4 and IL-13. Thus, differentiation from native helper T cells into either T_H1 or T_H2 also plays an important role in the production of IgE. Exposure to IL-4 favors the differentiation into T_H2 whereas T_H1 is favored in the presence of macrophage-derived cytokines such as IL-12, IL-18 and IFN-g.

ASTHMA

Asthma is caused by the obstruction of bronchial airway arising from inflammation and structural changes (remodeling) in the airway wall. Airway remodeling includes:

1. Airway wall thickening,
2. Subepithelial fibrosis,
3. Increased myocyte muscle mass,
4. Myofibroblast hyperplasia, and
5. Mucus metaplasia (review).

Asthma is often developed in childhood by interaction between environmental agents and genes that make bronchial epithelium susceptible to injury. Activation of Th2 cells releases IL13 and IL4, triggering inflammation and remodeling processes.

ASSOCIATED GENES

ADAM33

Encodes a disintegrin and metalloprotease domain 33. Polymorphisms of this gene are strongly associated with asthma, but the molecular mechanism remains to be elucidated.

ADRB2

Encodes beta-2-adrenergic receptor. Binding of its agonist stimulates cAMP formation, resulting in smooth muscle relaxation and bronchodilation (more info).

CCL11

Encodes a Cys-Cys cytokine gene that is involved in immunoregulatory and inflammatory processes.

CYSLTR1

Encodes cysteinyl leukotriene receptor which is involved in proliferation of smooth muscle and eosinophil migration.

FCER1A

Encodes an Fc fragment of IgE receptor, which plays a central role in allergic disease.

HLA-DQA1, -DQB1, -DRB1

Encodes α subunit of major histocompatibility complex.

IL13

Encodes interleukin 13.

IL4

Encodes interleukin 4.

IL5

Encodes interleukin 5.

MADH2

Encodes SMAD2 protein which is involved in TGF-b signaling.

MMP9

Encodes matrix metalloproteinase 9.

PHF11

Encodes a finger protein that could be involved in transcription.

SCGB1A1

Encodes Clara cell secretory protein CC16.

STAT6

Encodes STAT6 which plays the central role in mediating signals from IL4 and IL13.

TGFB1

Encodes transforming growth factor TGF-b, which plays an important role in fibrosis.

TNF

Encodes tumor necrosis factor.

ATHEROSCLEROSIS

Atherosclerosis is the buildup of plaque on the inside walls of arteries. Plaque is made up of low density lipoprotein (LDL), macrophages, smooth muscle cells, platelets, and other substances. It may narrow the lumen of a blood vessel and restrict blood flow. Plaque rupture can induce the formation of thrombus (blood clot) and block blood flow. This will result in ischemic stroke or heart attack.

Formation of Foam Cells

The first stage in the development of atherosclerosis is the formation of foam cells (macrophages with ingested oxidized LDL). The process begins with trap of LDL in the intima, which lies just below the endothelium (the monolayer of cells lining the arterial wall). Trapped LDL could be oxidized, triggering recruitment of monocytes into the intima. Several adhesion molecules are involved, including vascular-cell adhesion molecule (VCAM), integrin, selectin, and others. After entering the intima, monocytes differentiate into macrophages and ingest oxidized LDL.

Formation of Plaque

As atherosclerosis progresses, T lymphocytes, platelets and smooth muscle cells also join foam cells, expanding the plaque size. This involves cytokines to activate T lymphocytes and growth factors to promote proliferation of smooth muscle cells. Platelets can also release cytokines and growth factors to enhance migration and proliferation of smooth muscle cells. During this stage, a fibrous cap is formed to separate the plaque from the lumen.

Thrombosis

Thrombosis (formation of thrombus) arises from plaque rupture. Macrophages may release metalloproteinases and other proteolytic enzymes to degrade fibrous cap, making it susceptible to rupture. Plaque rupture activates platelets, leading to formation of blood clots at the site of lesion (mechanism from Internet Stroke Centre).

ASSOCIATED GENES

APOA1

Encodes apolipoprotein A-I, which is the major protein component of high density lipoprotein (HDL). Since HDL is less likely than LDL to enter the intima, high level of HDL is generally good for your health.

APOB

Encodes apolipoprotein B, which is the major protein component of LDL.

APOC2

Encodes a protein component of very low density lipoprotein (VLDL).

ABCA1

Encodes an ATP-binding cassette (ABC) transporter, which is involved in clearing excess cholesterol and generating HDL.

CCL2

Encodes the cytokine MCP-1 that has chemotactic activity for monocytes.

CETP

Encodes cholestery ester transfer protein. A deficiency of CETP is associated with increased HDL levels and decreased LDL levels (review).

CX3CR1

Encodes a chemokine receptor involved in cell adhesion.

MMP9

Encodes a matrix metalloproteinase, which is involved in plaque rupture (review, polymorphism).

OLR1

Encodes oxidized LDL receptor.

CARDIOMYOPATHY

Cardiomyopathy is defined as the disease of heart muscle associated with cardiac dysfunction. Based on morphological and functional characteristics, it is classified into four categories: hypertrophic, dilated, arrhythmogenic right ventricular, and restrictive. Their molecular mechanisms are discussed in a detailed review article, which also provides background information such as the structure of cardiac myocytes and the mechanism of contraction.

Associated Genes

Gene	Protein	Disease
ACTC	Cardiac actin	HCM, DCM
CLP	Cardiac LIM domain protein	HCM
Cypher/ZASP	LIM domain binding 3	DCM
DES	Desmin	DCM, RCM
DMD	Dystrophin	DCM
DSP	Desmoplakin	ARVC
JUP	Junction plakoglobin	ARVC
LMNA	Lamin A/C	DCM
MYBPC3	Myosin binding protein C	HCM, DCM
MYH7	b myosin heavy chain	HCM, DCM
MYL2	Myosin light chain 2	HCM
MYL3	Myosin light chain 3	HCM
PRKAG2	AMPK g subunit	HCM
RYR2	Cardiac ryanodine receptor	ARVC
SGCD	d-Sarcoglycan	DCM
TAZ	Tafazzin	DCM
TNNC1	Troponin C	HCM
TNNI3	Troponin I	HCM, RCM
TNNT2	Troponin T2	HCM, DCM
TPM1	Tropomyosin 1 (a)	HCM, DCM

TTN	Titin	HCM, DCM
VCL	vinculin	DCM

Abbreviations: HCM= Hypertrophic Cardiomyopathy; DCM= Dilated Cardiomyopathy;

ARVC=Arrhythmogenic Right Ventricular Cardiomyopathy; RCM= Restrictive Cardiomyopathy

EPILEPSY

Epilepsy is characterized by recurrent seizures, resulting from abnormal neuronal excitability or circuits. To date, more than 10 genes have been found to be associated with epilepsy. Most of them encode ion channels, which are known to play essential roles in nerve excitation.

There are many different epilepsy syndromes, which may be divided into two broad categories: generalized and partial. Generalized epilepsy begins simultaneously in both cerebral hemispheres whereas partial epilepsy begins in a localized area. Genes associated with the following syndromes have been identified.

GENERALIZED EPILEPSIES

Benign Familial Neonatal Convulsion (BFNC)

This syndrome begins in a few days after birth, and usually remits within a few weeks even without treatment. Genes associated: KCNQ2, KCNQ3 (voltage-gated potassium channel).

Childhood Absence Epilepsy (CAE)

This syndrome begins at the ages of 4 to 8 years. During absence seizures, patients stare and lose consciousness for a few seconds and then return immediately to normal. The syndrome is due to alteration in the circuitry between the thalamus and the cerebral cortex. Genes associated: CACNA1H (T-type calcium channel) and GABRG2 (GABA receptor channel).

Juvenile Myoclonic Epilepsy (JME)

This syndrome is characterized by myoclonic jerks and generalized tonic-clonic seizures. Gene associated: GABRA1 (GABA receptor channel).

Generalized Epilepsy with Febrile Seizures Plus (GEFS$^+$)

This syndrome encompasses a spectrum of childhood-onset phenotypes including febrile seizures (with fever), febrile seizures plus (without fever), partial epilepsies, myoclonic astatic epilepsy, and severe myoclonic epilepsy of infancy. Genes associated: GABRG2 (GABA receptor channel), SCN1A, SCN1B, SCN2A (voltage-gated sodium channel).

Severe Myoclonic Epilepsy of Infancy (SMEI)

This syndrome begins in infancy with prolonged clonic febrile and afebrile seizures. Genes associated: SCN1A (voltage-gated sodium channel).

X-linked Epilepsy

This type of epilepsy is caused by mutation of one or more genes in X chromosome. It is often associated with mental retardation. Genes associated: ARX, STK9, SLC6A8, PHF6.

PARTIAL EPILEPSIES

Autosomal Dominant Nocturnal Frontal Lobe Epilepsy (ADNFLE)

This syndrome is characterized by clusters of nocturnal motor seizures, lasting from a few seconds to five minutes. Genes associated: CHRNA4, CHRNB2 (nicotinic acetylcholine receptor channel)

Autosomal Dominant Partial Epilepsy with Auditory Features (ADPEAF)

This syndrome is characterized by infrequent partial seizures with or without auditory disturbances. Genes associated: LGI1. The cellular function of LGI1 is not known, but it shares some sequence homology with MASS1 gene, which is associated with an audiogenic epilepsy in a mouse model.

ESSENTIAL TREMOR

Tremor is defined as an involuntary, rhythmic oscillatory movement of a part or parts of the body. There are over 15 kinds of tremor. Essential tremor is the most common, about 10 to 20 times as prevalent as Parkinson's disease. It usually affects hands and head.

Essential tremor may arise from abnormal function of the olivocerebellar circuit. The precise mechanism, however, is not clear.

ASSOCIATED GENES

ETM1 (FET1)

Linkage analysis has mapped the gene to chromosome 3q13.1. However, the gene has not been identified.

ETM2

Linkage analysis has mapped the gene to chromosome 2p22-p25. It has not been identified yet.

Another gene mapped to chromosome 4p could also be involved.

HYPERTENSION

Blood pressure must be tightly regulated to permit proper supply of oxygen and nutrients to all vital organs. Low blood pressure indicates insufficient supply of oxygen and nutrients. High blood pressure (hypertension) delivers excess nutrients that may damage blood vessels, kidney and other organs. In addition, hypertension increases the workload on the heart, leading to early development of congestive heart failure. High pressure on blood vessels can also cause their rupture, resulting in stroke if it happens in the brain.

The blood pressure is equal to the cardiac output (blood volume pumped by the heart per minute) times the systemic vascular resistance. The two factors are regulated by complex systems including renin-angiotensin-aldosterone system, baroreceptors, natriuretic peptides, kinin—kallikrein system, adrenergic receptor system, and others. However, the major pathway leading to hypertension is surprisingly simple. It turns out that salt plays the central role in the pathogenesis of hypertension.

Major Pathway for Hypertension

When there is excess salt in the body, the blood volume increases, because the amount of water must increase to maintain plasma sodium concentration at a constant level. This is accomplished by two mechanisms:

1. Excess salt increases the osmolality of the body fluids, thereby stimulating the thirst centre to make the person drink more water.
2. The increase in osmolality also causes the release of antidiuretic hormone, which makes the kidneys to reabsorb water before it is excreted as urine. Since the blood pressure is proportional to the cardiac output (and the blood volume), this explains why excess salt should increase the blood pressure.

In addition to the direct relationship with blood pressure, cardiac output can also affect blood pressure indirectly through "autoregulation". Higher cardiac output triggers autoregulation mechanism to constrict blood vessels all over the body. Increased vasoconstriction also raises blood pressure.

Kidney and Salt

The amount of salt in the body is mainly controlled by kidneys, which filter over 170 liters of plasma containing 23 moles of sodium daily. More than 99 percent of the filtered sodium are reabsorbed at four locations:

1. The proximal tubule of the nephron (by Na^+/H^+ exchange),
2. The thick ascending loop of Henle (by Na-K-2Cl cotransport),
3. The distal convoluted tubule (by Na-Cl cotransport), and (4) the cortical collecting tubule (by epithelial sodium channels). Gene mutations that increase salt reabsorption will raise blood pressure.

The epithelial sodium channels at the last location are tightly regulated by the renin–angiotensin-aldosterone system. Renin is an enzyme that can convert angiotensinogen into angiotensin I. The latter can further be converted into angiotensin II by angiotensin converting enzyme (ACE). Angiotensin I is inactive but angiotensin II can stimulate the secretion of aldosterone, which then binds to the mineralocorticoid receptor, triggering a series of events that lead to the increase of epithelial sodium channel activity. When blood pressure or sodium level is low, renin is released to increase epithelial sodium channel activity, thereby increasing salt reabsorption.

ASSOCIATED GENES

ADD1

Encodes a cytoskeleton protein, adducin, which is involved in sodium reabsorption.

AGT

Encodes angiotensinogen, which can be converted into angiotensin by renin.

AGTR1, AGTR2

Encode angiotensin II receptors.

CLCNKB

Encodes a chloride channel that is involved in sodium reabsorption. Mutation of this gene causes Bartter syndrome type 3.

CYP11B2

Encodes a cytochrome P450 enzyme that is involved in the synthesis of aldosterone.

EDN1

Encodes endothelin-1, a powerful vasoconstrictor peptide produced by endothelial and smooth muscle cells.

GCGR

Encodes a glucagon receptor, which is involved in sodium reabsorption.

GNB3

Encodes a G-protein b3 subunit. The G protein is involved in signal transduction but the detailed mechanism is not known.

HSD11B1

Encodes the enzyme 11 b hydroxysteroid dehydrogenase, which catalyzes the conversion of cortisol to cortisone. Like aldosterone, cortisol can also activate mineralocorticoid receptor, but cortisone cannot. Therefore, the enzyme can protect mineralocorticoid receptor from being activated by cortisol.

KCNJ1 (ROMK)

Encodes a potassium channel that is involved in sodium reabsorption. Mutation of this gene causes Bartter syndrome type 2.

SLC12A1 (NKCC2)

Encodes a Na-K-2Cl cotransporter. Mutation of this gene causes Bartter syndrome type 1.

SLC12A3

Encodes a Na-Cl cotransporter. Mutation of this gene causes Gitelman syndrome.

SCNN1A,B,

Encodes the epithelial sodium channel. Mutation of either SCNN1B or SCNN1G causes Liddle syndrome.

WNK4

Encodes a serine-threonine kinase which regulates the balance between sodium reabsorption and potassium secretion.

LONG QT SYNDROME

The long QT syndrome is characterized by prolongation of the QT interval on electrocardiograms (ECG) which manifest the membrane potential changes of cardiac myocytes (more info). Seven types of long QT syndromes have been identified, designated as LQT-1 to LQT-7, each resulting from mutations of a particular gene. Since the membrane potential is determined by ionic fluxes through ion channels, most associated genes encode ion channels.

ASSOCIATED GENES

Syndrome	Gene	Protein
LQT-1	KCNQ1	voltage gated K^+ channel
LQT-2	KCNH2	voltage gated K^+ channel
LQT-3	SCN5A	voltage gated Na^+ channel a subunit
LQT-4	ANKB	Ankyrin-B
LQT-5	KCNE1	voltage gated K^+ channel
LQT-6	KCNE2	voltage gated K^+ channel
LQT-7	KCNJ2	voltage gated K^+ channel

The prolongation of the QT interval is due to slow repolarization of the membrane potential. Mutations in these K^+ channels cause the loss of function, thereby reducing outward K^+ current and consequently slowing repolarization. Mutations of the Na^+ channel, SCN5A, cause incomplete inactivation, thereby increasing inward Na^+ current which also slows repolarization.

The protein ankyrin-B responsible for LQT-4 is not an ion channel. However, it is an adaptor protein that binds to sodium pump, sodium/calcium exchanger, and ion channels.

OSTEOARTHRITIS

Osteoarthritis is characterized by degradation of articular cartilage (the softer parts of bones) and the formation of new bone (known as osteophytes) at the margins of the joints.

ASSOCIATED GENES

AGC1

Encodes aggrecan 1 which can withstand compression in cartilage.

COL1A1

Encodes the alpha-1 chain of type I collagen.

COL2A1

Encodes the alpha-1 chain of type II collagen.

COL9A1

Encodes one of the three alpha chains of type IX collagen.

COL11A2

Encodes one of the two alpha chains of type XI collagen.

ESR1

Encodes estrogen receptor 1 (alpha).

FRZB

Encodes frizzled-related protein

IGF1

Encodes insulin-like growth factor 1.

IL1R

Encodes interleukin 1 receptor.

MATN1

Encodes cartilage matrix protein.

TGFB1

Encodes transforming growth factor, beta 1.

TIMP3

Encodes tissue inhibitor of metalloproteinase 3.

VDR

Encodes vitamin D receptor.

OSTEOPOROSIS

Osteoporosis is characterized by reduction in bone mass and structural deterioration of bone tissue with a high risk of fractures. The bone mass depends on the balance between bone formation by osteoblasts and bone resorption by osteoclasts. Osteoblasts are derived from mesenchyme cells under the regulations of several transcription factors such as Cbfa-1, Osterix, Krox-20, Fra-1, Dlx5 and Twist. Differentiation of osteoclasts from the monocyte/macrophage lineage involves estrogen and RANK.

ASSOCIATED GENES

AHSG

Encodes alpha-2-HS-glycoprotein.

AR

Encodes androgen receptor.

BGLAP

Encodes bone gamma-carboxyglutamate protein (osteocalcin).

CALCA

Encodes calcitonin.

CALCR

Encodes calcitonin receptor.

CASR

Encodes calcium-sensing receptor.

COL1A1

Encodes the alpha-1 chain of type I collagen.

COL1A2

Encodes the alpha-2 chain of type I collagen.

CYP19

Encodes cytochrome P450, family 19.

ESR1

Encodes estrogen receptor 1 (alpha).

ESR2

Encodes estrogen receptor 2 (beta).

FOSL1 (FRA-1)

Encodes FOS-like antigen 1.

IGF1

Encodes insulin-like growth factor 1.

IL6

Encodes interleukin-6.

IL1RN

Encodes interleukin 1 receptor antagonist.

KL

Encodes klotho protein.

MTHFR

Encodes methylenetetrahydrofolate reductase (NADPH).

NR3C1

Encodes glucocorticoid receptor.

OPG

Encodes osteoprotegerin (tumor necrosis factor receptor superfamily, member 11b).

PPARG

Encodes peroxisome proliferative activated receptor, gamma.

PTH

Encodes parathyroid hormone.

PTHR1

Encodes parathyroid hormone receptor 1.

RUNX2

Encodes runt-related transcription factor 2.

TGFB1

Encodes transforming growth factor, beta 1.

TNF

Encodes tumor necrosis factor.

TNFR2

Encodes tumor necrosis factor receptor 2.

VDR

Encodes vitamin D receptor.

WRN

Encodes Werner Syndrome helicase.

PARKINSON'S DISEASE

Parkinson's disease is characterized by the presence of Lewy bodies and the loss of dopamine-producing neurons in substantia nigra that controls muscle movement. The Lewy body is an abnormal structure found in certain areas of the brain. It contains a protein called α-synuclein, which plays the central role in Parkinson's disease and other diseases involving Lewy bodies, such as dementia with Lewy bodies, multiple system atrophy, and Hallervorden-Spatz disease.

α-synuclein

α-synuclein is a 140 amino acid protein abundantly expressed in presynaptic terminals of vertebrates. One of its normal functions is to regulate dopamine transporter activities. This protein contains an NAC region that is prone to aggregate, especially under oxidative stress (1, 2, 3). The aggregated α-synuclein can inhibit the function of 26S proteasome, which is important for the clearance of misfolded proteins and other target molecules. The dysfunction of proteasome will contribute to cell death.

Two mutations, A53T and A30P, in α-synuclein have been identified in families with early-onset familial Parkinson's disease. These mutations may accelerate the aggregation of α-synuclein. It is also interesting to note that, even without mutation, extra copies of the gene encoding α-synuclein can cause Parkinson's disease at an average age of just 34.

α-synuclein knockout mice are still viable and exhibit normal dopaminergic cell bodies and synapses. In the absence of α-synuclein, mice are resistant to MPTP-induced toxicity. MPTP is known to inhibit mitochondrial complex I, resulting in Parkinson-like symptoms. Another complex I inhibitor, rotenone, produces more widespread neurotoxicity when injected into rats. Removing α-synuclein did not protect the rotenone-induced toxicity. It is known that entry of MPTP into the dopamine neuron is mediated by dopamine transporter, whereas rotenone is highly lipophilic and may enter the neuron directly.

Parkin

Parkin is a 465 amino acid protein involved in the ubiquitin-proteasome system for the clearance of misfolded proteins and other target molecules. It serves as the E3 enzyme in the system. Mutations of the Parkin gene are associated with early onset Parkinson's disease.

Parkin does not seem to interact with native α-synuclein, but it does ubiquitinate an α-synuclein-interacting protein, synphilin-1. α-synuclein, Parkin and synphilin-1 represent the major components of Lewy bodies. In the Lewy body, α-synuclein is also ubiquitinated, possibly by Parkin. It seems that Parkin can protect the cell by specifically tagging modified (oxidized, aggregated, etc.) α-synuclein for destruction.Whether synphilin-1 plays a role in the recognition of modified α-synuclein remains to be investigated.

DJ-1

Mutations in DJ-1 have been reported in families with early-onset Parkinson's disease. The function of this protein is not clear. Its structure resembles a cysteine protease. Dimerization is required for its catalytic activity. The disease-associated L166P mutation may disrupt the dimerization (1, 2).

GSTO1

GSTO1 is involved in both Parkinson's disease and Alzheimer's disease. The role of GSTO1 in these diseases is not clear. However, it is known that GSTO1 can modulate ryanodine receptors, which are calcium channels in the endoplasmic reticulum of various cells, including neurons. Ca^{2+} ions play an important role in neurotransmitter release. They also bind to α-synuclein to regulate its function. On the other hand, Ca^{2+} ions may exercise pathological effects by promoting oligomerization of α-synuclein

RHEUMATOID ARTHRITIS

Rheumatoid arthritis is characterized by chronic inflammation of the synovium (the lining of the joints), which often leads to destruction of articular cartilage and bone. The disease is caused by complex cell-cell interactions, resulting in the release of various cytokines such as TNF?, IL-1, RANKL (receptor activator of NF-kB ligand) and IL-17. These cytokines may damage articular cartilage and bone.

ASSOCIATED GENES

HLA-DRB1

Encodes major histocompatibility complex, class II, DR beta 1. The role of this protein in rheumatoid arthritis may not be related to binding and presentation of arthritogenic peptides because its crystal structure indicates that the susceptibility epitope residues face away from its antigen-bindig groove. Recent evidence suggests that it may be related to premature telomeric loss.

TNF

Encodes tumor necrosis factor.

TNFR1

Encodes TNF receptor member 1A.

TNFR2

Encodes TNF receptor member 1B.

IFNG

Encodes Interferon gamma.

CTLA4

Encodes cytotoxic T lymphocyte-associated protein 4.

IKBL

Encodes IkB-like protein.

MIF

Encodes macrophage migration inhibitory factor.

PTPRC (CD45)

Encodes protein-tyrosine phosphatase receptor type C.

Chapter 10

Molecular Mechanisms of Learning and Memory

INTRODUCTION

Albert Einstein said: "the most incomprehensible fact about nature is that it is comprehensible". Indeed, physical phenomena are comprehensible. Weather changes, chemical reactions and molecular structures appear quite complex, but their underlying principles are amazingly simple. All physical phenomena are governed by four fundamental forces - gravity, electromagnetism, strong and weak nuclear forces. Recently, these forces can further be unified by a single theory — the string theory.

What about the brain? Is its ability of comprehension comprehensible? In the seventeenth century, Isaac Newton had discovered the gravitational law and Robert Boyle had formulated the law for gas property, but human beings were still quite ignorant about physiological systems. William Harvey, a pioneer in blood circulation, told Robert Boyle that it was the valves in the veins which guided him to the next phase of his thinking.

We now know that thoughts are not controlled by blood vessels and sorrow has nothing to do with a "broken heart". They come from our brain — the only thing which can try to understand itself. With a size smaller than a TV set, our brain contains more than 10 billions of neurons. Each neuron is connected to many (up to thousands) other neurons. The junction between neurons is called the synapse. Brain functions are the results of signal transmission through trillions of synapses.

Since neurons are made up of the same atoms as found in the physical world, it might be possible to explain the brain functions in terms of physical laws. In the past few decades, significant progress has been made toward this goal. The mechanism of learning can be explained at the molecular level. We shall start from the basic information, namely, the generation of nerve impulses and the function of ion channels. You may jump to any topic on your left.

NERVE IMPULSES

Nervous systems, like other communication systems, use a sequence of impulses to carry message. The nature of nerve impulses, however, differs entirely from electromagnetic waves and sound waves. In every nerve cell, there is a membrane separating the cytoplasmic fluid from the extracellular solution. The transmembrane voltage (also called membrane potential) is defined as the inside potential minus the extracellular potential. When the nerve fiber is at rest, the membrane potential is about -70 mV. The nerve impulse is a sharp change of the membrane potential. Therefore, it is also known as the action potential.

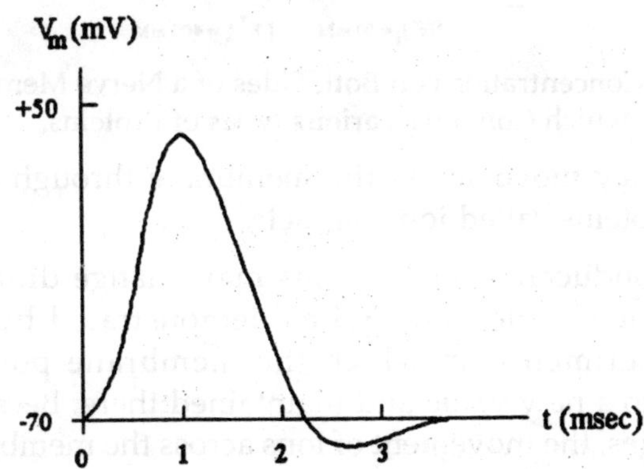

Fig. A Typical Nerve Impulse (Action Potential).

Ion Conductances

The generation of action potentials is mainly due to the changes of sodium (Na^+) and potassium (K^+) conductances. Figure shows the concentrations of Na^+ and K^+ ions on both sides of a nerve membrane. For Na^+, its concentration on the extracellular side is much higher than inside. We immediately notice that Na^+ ions are far from electrochemical equilibrium — both the electric force due to electric potential difference and the chemical force due to ion concentration difference are pointing inward. How could the nerve membrane maintain such a stable state? This is because the conductance of Na^+ ions in the membrane is very small at the resting membrane potential. Although the inward driving force is large, the resulting Na^+ influx is small. This small influx can be balanced by a slow ion transport process, the Na^+-K^+ pump, which moves the Na^+ ions outward and simultaneously the K^+ ions inward.

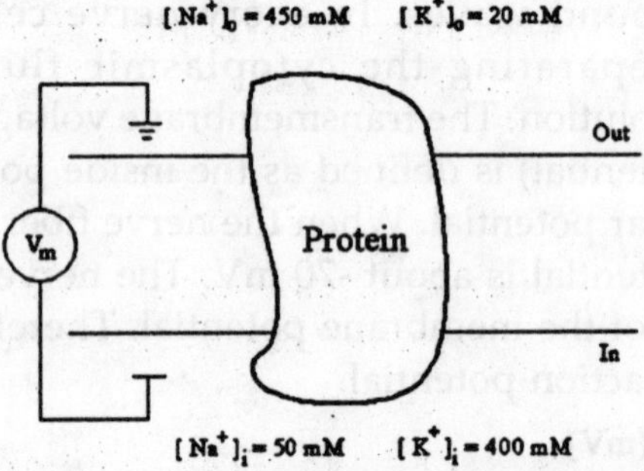

Fig. Ion Concentrations on Both Sides of a Nerve Membrane, which Contains Various types of Proteins.

Ions may move across the membrane through a special class of proteins called ion channels.

The conductance of Na^+ ions may change dramatically with the membrane potential as demonstrated by voltage clamp experiments, in which the membrane potential is displaced to a new value and maintained there. Because ions carry charges, the movement of ions across the membrane will change the membrane potential. To maintain a constant membrane potential, the voltage clamp circuit must generate electric currents to neutralize the membrane potential change caused by the ionic flux. Thus, the ion current through the membrane is reflected in the electric current of the voltage clamp circuit outside the membrane.

A membrane is said to be depolarized if the new potential V_m is more positive than the resting potential $V_m{}^o$, and hyperpolarized if $V_m < V_{mo}$. Since depolarization makes the inside potential more positive, we expect (from the consideration of electrochemical forces) the currents of all cations to increase in the outward direction. Figure shows the experimental result for squid axons. The K^+ current indeed increases in the outward direction. However, the behaviour of Na^+ current is totally unexpected. It first rises sharply in the inward direction and then declines with a slower rate to its resting value. Phenomenologically, the change of the Na^+ current can be attributed to the change in Na^+ conductance. Experimental results indicate that the Na^+ conductance increases in the early stage of depolarization, which exceeds the reduction in the inward driving force. As time proceeds, the Na^+ conductance decreases, resulting in the late decline of Na^+ currents. The rise and fall of Na^+ conductance are known as sodium activation and sodium inactivation, respectively. They play important roles in the generation of a nerve "impulse", which must contain a rising phase as well as a falling phase. Mathematically, the ion conductance is defined by

$$G_{ion} = I_{ion}/(V_m - V_{ion}) \qquad (1)$$

where I_{ion} denotes the ion current through the membrane (outward current is defined as positive), and V_{ion} is the equilibrium potential for the ion. According to the Nernst equation,

$$V_{ion} = (RT/F) \ln ([ion]_o/[ion]_i) \qquad (2)$$

where R denotes the gas constant, T is the absolute temperature, F represents the Faraday constant, $[ion]_o$ and $[ion]_i$ are the ion concentrations in the extracellular and intracellular fluids, respectively. At room temperature, the equilibrium potential for Na^+ with concentrations given in Fig. When the membrane potential is depolarized, I_{ion} varies with time as shown in Fig.. From eq.(1), we can obtain the time course of the ion conductance. Fig. shows the experimental results for a few depolarizations. We see that the peak of Na^+ conductance increases with increasing depolarizations before it levels off at large depolarizations. Another feature to remember is that the rising rate of G_K is much slower than G_{Na}.

The ion conductance change is the collective results of many individual ion channels. Each ion channel may be in either "open" or "closed" state. The opening probability of each ion channel depends on the membrane potential, but the detailed mechanism is not known.

THE GENERATION OF ACTION POTENTIALS

In excitable cells, the action potential is elicited when the membrane potential is depolarized to a critical value (the threshold). In most axons, the threshold is about 15 mV above the resting potential. Before the threshold is reached, the axon membrane behaves like an ordinary inactive substance with a certain resistance and capacitance. When a constant current I_s is applied to the inactive membrane, the membrane potential becomes,

$$V_m(t) = I_s R_m[1 - \exp(-t/R_m C_m)] + V_{mo} \quad (3)$$

where R_m and C_m are the effective resistance and capacitance of the membrane, respectively.

As shown in Fig. the peak of G_{Na} increases with increasing depolarizations. When the membrane is depolarized, more Na^+ ions will flow into the intracellular fluid. Because the Na^+ ions carry positive charges, the Na^+ influx will make the membrane more depolarized, which in turn makes the Na^+ conductance even larger. In the mean time, the K^+ conductance also increases with increasing depolarization. In the physiological range of membrane potentials and ion concentrations, the electrochemical force on K^+ is outward. The outflux of K^+ will make the membrane more hyperpolarized, counteracting the Na^+ influx. However, the rising rate of G_K is much slower than G_{Na}. In the early stage, the rise of K^+ outflux will not be as large as the Na^+ influx except at small depolarizations. Therefore, when the membrane is depolarized to a critical value, above which the Na^+ influx exceeds the K^+ outflux, the Na^+ ions will flow inward at an accelerating rate. As a result, the membrane potential rises sharply until it approaches the equilibrium potential for Na^+. Subsequently, the sodium inactivation sets in and the K^+ conductance rises significantly. The membrane potential then returns to its resting value, constituting the falling phase of the action potential.

If the stimulus fails to raise the membrane potential to the threshold, the intrinsic sharp potential change cannot occur. This behaviour is known as "all or none". Figure illustrates the membrane response to two different current pulses, I_1 and I_2. The magnitude of I_1 is too small to charge the membrane above the threshold, but I_2 is strong enough to initiate an action potential.

ION CHANNELS

Ion channels are a special class of proteins that conduct small ions such as Na^+, K^+, Ca^{2+} or Cl^-. The opening and closing of ion channels may depend on membrane voltage or the binding of molecules (the ligand). Its mechanism remains unclear. However, the mystery of their selectivity has been solved.

ION SELECTIVITY

Potassium Channels

By definition, a potassium channel conducts mainly K^+ ions while excluding Na^+ and Ca^{2+} ions. We note that a Na^+ ion is smaller than a K^+ ion. How could a channel conduct K^+ ions while excluding smaller ions? The answer lies in the ion's hydration energy.

The diameter of the potassium channel's selectivity filter (the narrowest region of its channel pore) should be about the same as the size of a Cs^+ ion (diameter = 3.3Å) because Cs^+ is the largest ion which can barely pass through the potassium channel. In a solution, ions are surrounded by water molecules. The first hydration shell of a K^+ or Na^+ ion contains six water molecules. The diameters of unhydrated Na^+ and K^+ ions are 1.96Å and 2.66Å, respectively. The effective diameter of a water molecule in a hydrated ion is larger than 2Å. Therefore, to pass through the selectivity filter, the K^+ or Na^+ ion must remove four water molecules from its first hydration shell, leaving only two (one at the front and anther at the back). The smaller Na^+ ion requires greater dehydration energy than the K^+ ion, because its nucleus has shorter distance with surrounding water molecules and thus interacting more strongly. As a result, the Na^+ ion is harder to pass through the potassium channel than the K^+ ion. The observation that lithium (Li^+) is excluded from the potassium channel can also be explained by the same energetic consideration. Divalent cations such as Ca^{2+} should also be excluded because their dehydration energies are much higher than monovalent cations.

Sodium Channels

In the sodium channel, the Na^+ ion is more permeable than the K^+ ion. This is because the selectivity filter of the sodium channel is slightly larger than that of the potassium channel. It is large enough to accommodate a Na^+ ion attached with three water molecules, but not enough for a K^+ ion attached with three water molecules. Therefore, to pass through the sodium channel, the Na^+ ion needs to remove only three, but the K^+ ion has to remove four, water molecules from its first hydration shell. The required dehydration energy for the K^+ ion is greater than the Na^+ ion.

Calcium Channels

In calcium channels, the permeability of monovalent cations (Na^+ and K^+) is about three orders of magnitude smaller than the Ca^{2+} permeability. This ion selectivity does not seem to involve hydration, because Ca^{2+} is more heavily hydrated than Na^+, and the unhydrated diameters of Ca^{2+} and Na^+ are almost identical. Then, how could calcium channels select Ca^{2+} over Na^+?

Although the permeability of monovalent cations in the calcium channel is quite small at normal ionic concentrations, large monovalent cationic current can be observed in the absence of Ca^{2+} and other divalent cations. This suggests that the calcium channel is basically permeable to both divalent and monovalent cations, but the selectivity arises from competition between ions. The calcium channel may contain a negatively charged binding site to facilitate ion conduction. The monovalent cations simply cannot compete with Ca^{2+} for this binding site. This idea has been demonstrated experimentally. In the calcium channel, if a negatively charged glutamate residue in the pore-lining region is mutated into a positively charged lysine, the calcium channel becomes more permeable to Na^+ than Ba^{2+}. Conversely, in the sodium channel, mutation of a pore-lining lysine residue into glutamate transforms the channel from a Na^+-selective to a Ca^{2+}-selective channel.

CHANNEL STRUCTURES

The detailed 3D structures of most ion channels are not known. However, from the hydrophobicity profile of their amino acid sequences, it is possible to obtain the domain structures of most ion channels.

K Channels

There are many types of potassium channels. Most commonly observed K channels are composed of four subunits, each is homologous to the Shaker protein. The hydrophobicity profile indicates that it contains six hydrophobic segments, designated as S1 - S6. These segments are likely to be the transmembrane domains. Other experimental results suggests that the P-region is lining the channel pore.

Na and Ca Channels

A Na or Ca channel consists of a major pore-forming subunit and possibly other small auxiliary subunits. The major pore-forming subunit is called the α subunit which can be divided into four similar domains. Each domain is analogous to a Shaker protein with six transmembrane segments and a P-region. Thus, an a subunit is sufficient to form an ion channel.

SYNAPTIC TRANSMISSION

A neuron is composed of three parts: dendrites, axon and nerve terminals. The dendrites receive signals from other neurons, the axon conducts action potentials, and the nerve terminals transmit signals to other neurons. At the dendrite, the action potential can be elicited if the membrane potential is depolarized to a threshold. The action potential then travels along the axon to the nerve terminal, where depolarization may open calcium channels for the entry of Ca^{2+} ions, which then induce the release of neurotransmitters stored in the vesicles. In a neural network, the nerve terminal of a neuron may form a synapse with either the dendrite or nerve terminal of another neuron. The synaptic cleft between two neurons is about 200 - 500 Å wide. Neurotransmitters released from the vesicle may diffuse through the synaptic cleft to act on their receptors in the postsynaptic neuron.

The receptors for neurotransmitters can be classified into two categories: G-protein-coupled receptors and ionotropic receptors. G-protein-coupled receptors are involved in signal transduction. Like other agonists, binding of neurotransmitters on G-protein-coupled receptors may trigger a series of signaling processes (more info). The ionotropic receptors form an ion channel which may be activated upon binding of specific neurotransmitters. Once activated, the influx of cations (e.g., Na^+) may cause the postsynaptic membrane to depolarize. If the depolarization reaches the threshold, an action potential can be generated on the postsynaptic neuron.

The ion channel formed by neurotransmitter receptors is called the synaptic channel. Each synaptic channel consists of five receptors (subunits). Figure shows the general structure of a nicotinic acetylcholine receptor (nAChR) channel.

The glutamate receptor (GluR) channel plays a critical role in the Hebbian type of learning. It may be divided into three subtypes: NMDA, kainate and AMPA. All glutamate receptors bind efficiently to the glutamate molecule, but their binding affinity for other agonists varies. The NMDA subtype has a high binding affinity for the NMDA molecule, but low affinity for the kainate or AMPA molecule. Therefore, a channel formed by NMDA receptors can be activated by NMDA or glutamate, but not by kainate or AMPA. Similarly, the channel formed by AMPA receptors can be activated by AMPA or glutamate, but not by NMDA or kainate.

LEARNING - OVERVIEW

Learning, as recognized by Ivan Pavlov, is basically a process which associates two events. After we see rain following dark cloud a number of times, the dark cloud will very likely trigger the memory of rain before it actually appears. The association between rain and dark cloud allows us to avoid unpleasant events well ahead. This ability is important for survival. When one learns to play chess, he should first memorize the patterns which will immediately lead to the capture of the king. Let us call these patterns {A}. After a few weeks of experience (learning), he may find that some patterns, called {B}, could lead to {A}. The association between {A} and {B} also allows the player to better plan his next move before too late. After years of experiences, he may discover that some patterns {D} are likely to result in patterns {C}, and {C} may lead to {B}. Therefore, a novice player can see only a couple of steps ahead, but an experienced player can see many steps in advance.

A major goal of physical sciences is to find the equations associating the state of a system at a given time t and its next moment, $t + dt$. This enables physicists to predict future time courses, given the present information (an initial condition). For instance, meteorologists have been able to develop a set of equations describing the time courses of atmospheric conditions (temperature, pressure, etc.). From the present information provided by satellites, they can predict rains several days before the dark cloud is formed.

In our brain, the *hippocampus* is the major area for learning. In 1950s, a few patients were treated for epilepsy by removing part of the hippocampus. After the surgery, the patients still remembered their names, the events of their childhood, and the language learned before the surgery. However, they could not form new memory. Although they could talk normally with their relatives and friends, but they forgot such events in the next day. The early stage of the Alzheimer's disease is characterized by the same symptom, because the hippocampus is usually the first area to be damaged in Alzheimer's patients.

At the cellular and molecular level, there are two distinct learning mechanisms: the Pavlovian type and Hebbian type. In the Hebbian type of learning, both presynaptic and postsynaptic neurons are involved. The Pavlovian type does not involve postsynaptic neurons. Learning in the hippocampus is mainly the Hebbian type.

THE PAVLOVIAN CONDITIONING

The Pavlovian conditioning is an implicit learning which does not require consciousness of the subject. Hippocampus-impaired patients can still learn new motor skills through this mechanism. Even invertebrate animals with simple neural circuits may be trained by the Pavlovian conditional reflex, which is essentially the association between conditioned and unconditioned stimuli. In the original Pavlov's experiments on the dog, food is the unconditioned stimulus which elicits the secretion of saliva by native response. The unconditioned stimulus was preceded by a conditioned stimulus (the bell-ringing) which initially did not produce salivation. After a few paired stimulations, the bell-ringing itself could produce salivation.

The molecular mechanism of the Pavlovian conditioning has been well understood for the Aplysia system which contains a simple neural circuit. In this system, the unconditioned stimulus is a strong shock to the tail, eliciting gill movement by native reaction. The conditioned stimulus is a weak tactile stimulus to either siphon or mantle shelf, which initially does not cause the gill movement. After a few paired stimulations (weak stimulus to siphon followed by tail shock), the weak stimulus to siphon alone can cause the gill movement. To explain its molecular mechanism, let us first explain the mechanism of sensitization, which refers to the potentiation of synaptic transmission by the unconditioned stimulus alone.

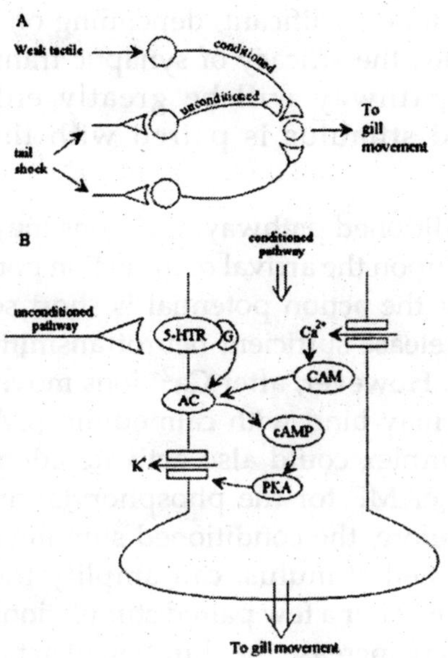

Fig. Molecular Mechanism of the Pavlovian Conditioning for the Aplysia System.

In Fig. the unconditioned stimulus (tail shock) not only elicits gill movement directly, but also activates a modulatory neuron which releases 5-HT (serotonin) to bind on its G-protein-coupled receptor (5-HTR) in the nerve terminal of the conditioned pathway. This nerve terminal may release neurotransmitters to excite a postsynaptic neuron which is directly responsible for the gill movement. Binding of 5-HT to its receptor can activate adenylyl cyclase (AC), producing cAMP. The cAMP triggers phosphorylation of potassium channels by activating the protein kinase A (PKA). Phosphorylation of potassium channels in this system reduces channel opening probability. The reduction of K^+ currents slows the repolarization of action potentials, thereby increasing

Ca^{2+} influx to release more excitatory neurotransmitters. Hence, the synaptic transmission in the conditioned pathway is potentiated.

The sensitization by the unconditioned stimulus alone may or may not be significant, depending on the system. As explained below, the efficacy of synaptic transmission in the conditioned pathway will be greatly enhanced if the unconditioned stimulus is paired with the conditioned stimulus.

In the conditioned pathway, Ca^{2+} ions may move into the nerve terminal upon the arrival of the action potential. Initially, the duration of the action potential is short so that the Ca^{2+} influx cannot release sufficient neurotransmitters to elicit the gill movement. However, after Ca^{2+} ions move into the nerve terminal, they may bind with calmodulin (CAM). The Ca^{2+}/calmodulin complex could also activate adenylyl cyclase to produce more cAMP for the phosphorylation of potassium channels. Therefore, the conditioned stimulus, if paired with the unconditioned stimulus, can amplify the activation of adenylyl cyclase. After a few paired stimulations, the reduction of K^+ outflux may increase the duration of action potentials to the extent that the conditioned stimulus (i.e., the weak stimulus to siphon) alone can induce sufficient Ca^{2+} influx to elicit the gill movement.

In the Pavlovian conditioning, the unconditioned stimulus has to be applied shortly after the conditioned stimulus. The optimal interval between the two stimuli was found to be 0.5 second for the case of Aplysia. The 0.5 second interval may correspond to the time required for the binding of Ca^{2+} with calmodulin. Thus, the adenylyl cyclase can be activated almost simultaneously by the Ca^{2+}/calmodulin complex from the conditioned pathway and the G protein from the unconditioned pathway.

HEBBIAN TYPE OF LEARNING

As discussed in the previous page, the Pavlovian type of learning involves only the presynaptic neuron. In the Hebbian type of learning, the synaptic modification is induced by the participation of both presynaptic and postsynaptic neurons.

Recalling that learning is the ability to associate two events that happen almost at the same time. In the hippocampus, events are represented by a population of neurons, each may be either excited or in the resting state. A particular event is represented by a particular set of neurons in the excited state. For instance, if the number of neurons involved in the representation is *n*, then mathematically an event can be denoted by a vector with the dimension *n*,

$$X = [x_1, x_2, x_3, \ldots x_n]$$

where x_i ($i = 1 - n$) is either "0" (resting) or "1" (excited).

The hippocampus contains the most complex neural network. Each neuron is connected to thousands of other neurons. For simplicity, we assume that only five neurons are involved in the representation of events. The connection of these neurons are shown in Figure. The lines are drawn from the nerve terminals (presynaptic) in the upper row to the dendrites (postsynaptic) in the lower row. For example the terminals of the first neuron are connected to the dendrites of the second and fourth neurons. The terminals of the second neuron are connected to the dendrites of the first and third neurons. These lines may also be considered as the synapses between neurons. The "dark cloud" is represented by the excitation of the second and fourth neurons and the "rain" is represented by the excitation of the first, fourth and fifth neurons.

In the Hebbian type of learning, the synaptic modification may be induced only when both presynaptic and postsynaptic neurons are excited. In Figure these synapses are represented by red lines. The left red line connects the second and first neurons; the right red line connects the fourth and fifth neurons. When the dark cloud and rain happen almost at the same time, these neurons are excited and their synapses are modified so that nerve impulses can be more easily transmitted from the presynaptic neuron to the postsynaptic neuron.

Suppose before learning the nerve impulse is unlikely to transmit from the the second neuron to the first neuron. After the pairing between dark cloud and rain, this synaptic transmission is greatly enhanced. Next time, the dark cloud alone is likely to cause the excitation of the first neuron, thereby increasing the probability to recall the rain.

How could the physiological system implement the Hebbian type of learning - namely, excitation of both presynaptic and postsynaptic neurons induces synaptic modification? The answer is the NMDA channel, which is a subtype of glutamate receptor channels. For most synaptic ion channels, activation (opening) requires only the binding of neurotransmitters. However, activation of the NMDA channel requires two events: binding of glutamate (a neurotransmitter) and relief of Mg^{2+} block. NMDA channels are located at the postsynaptic membrane. When the membrane potential is at rest, the NMDA channels are blocked by the Mg^{2+} ions. If the membrane potential is depolarized due to excitation of the postsynaptic neuron, the outward depolarizing field may repel Mg^{2+} out of the channel pore. On the other hand, binding of glutamate may open the gate of NMDA channels (the gating mechanisms of most ion channels are not known). In the normal physiological process, glutamate is released from the presynaptic terminal when the presynaptic neuron is excited. Relief of Mg^{2+} block is due to excitation of the postsynaptic neuron. Therefore, excitation of both presynaptic and postsynaptic neurons may open the NMDA channels.

Another important feature of the NMDA channel is that it conducts mainly the Ca^{2+} ion which may activate various enzymes for synaptic modification. The enhancement of synaptic transmission is called the long-term potentiation (LTP), which involves two parts: the induction and the maintenance. The induction refers to the process which opens NMDA channels for the entry of Ca^{2+} ions into the postsynaptic neuron. The subsequent synaptic modification by Ca^{2+} ions is referred to as the maintenance of LTP. We have just explained the mechanism for the induction of LTP. Mechanisms for the maintenance of LTP have not been fully understood. However, it is known that the AMPA receptor trafficking plays an important role.

The AMPA receptor is a subtype of glutamate receptor. They may form ion channels to conduct small cations. Their density at the postsynaptic membrane can regulate synaptic strength. Higher density will allow greater ionic influx upon binding of glutamate released from the presynaptic terminal, thereby resulting in more membrane depolarization, which in turn facilitates transmission of the nerve impulse from the presynaptic neuron to the postsynaptic neuron. For this reason, AMPA receptors are continuously being inserted into or removed from the postsynaptic membrane. This AMPA receptor trafficking is controlled by the Ca^{2+} ion.

HOLOGRAPHIC MEMORY

The long term potentiation in the hippocampus is a temporary memory, lasting for a few hours to a couple of days. It is equivalent to the working memory (RAM) of a personal computer. After the computer is shut off, the working memory is gone, unless the contents have been saved in a hard disk.

Some of the memory acquired in the hippocampus will be transferred to the cortex for permanent storage. This involves synthesis of new proteins and possibly formation of new synapses. The detailed mechanism is not clear.

In the 1920s, Karl Lashley was trying to locate the exact area for memory storage. He trained rats to perform specific tasks, then impaired specific areas of the cortex. He found that the memories were reduced, but not removed by the lesions at any part of the cortex. This suggested to him that the memory was not stored at any specific area, but rather distributed over the entire cortex. In 1960s, Karl Pribram, who worked with Karl Lashley in the 1940s, proposed that the memory was stored as a hologram.

The hologram was invented in 1947 by Dennis Gabor to record images in a medium (e.g., a crystal). It has a very unique feature, namely, any small part of the medium contains all information about the image. Therefore, even if the medium is broken into pieces, the entire image can still be obtained from any piece, except the resolution is reduced.

The holographic storage of memory has at least the following advantages:

1. Resistance to damage: If some parts of the medium are damaged, all information can still be obtained from other parts.
2. Efficient retrieval: All information can be retrieved from any part of the medium.

Because of these advantages, the brain memory may indeed be stored as a hologram. For physical memory disks such as DVD, the present technique has reached the upper limit of data density. Computer engineers are trying to use hologram for the next generation of data storage, because it may store significantly higher amount of information. In our brain, the holographic memory may also be able to store more information than other alternatives.

COMPREHENSION

Everyone can associate rain with dark cloud, but it requires Newton's ingenuity to associate the fall of apples with moon's motion, and Einstein's imagination to relate energy with mass by the equation, $E = mc^2$. As a matter of fact, most people possess the ability similar to what Newton and Einstein had: forming a theory to associate (explain) events already experienced by the brain. This ability is not limited to researchers in physical or biological sciences. A stock market trader, after years of experiences, can also form a theory to predict the up and down of the market. Of course, not every theory is correct. When a set of events can be associated by a theory, these events will become "reasonable" or "comprehensible" to the brain. If the theory is indeed correct, it will have the power to predict future events, like weather forecast.

The goal of biophysics is to associate biological phenomena with physical laws. Although many theoretical gaps remain to be filled, there are good reasons to believe that all biological processes can eventually be explained in terms of the properties of atoms and molecules. In physical chemistry, the properties of atoms and molecules have been able to be explained by quantum mechanics. In theoretical physics, an even more fundamental theory has been developed, which can unify all four fundamental forces (gravitational, electromagnetic, strong and weak nuclear forces). This "theory of everything" is known as the string theory.

A revolutionary discovery of the string theory is that the whole universe (called the "bulk universe") should have ten dimensions of space, instead of three dimensions (length, width and height). The reason why we perceive only three dimensions of space is because our matter and electromagnetic waves (photons) are confined in a sub-universe called the "braneworld". It is impossible for us to see anything outside our braneworld by light or other electromagnetic waves (more info). Imaging that there is a creature confined on a TV screen. He thought his world was two dimensional because he cannot see anything outside the TV screen. He does not know there is a three dimensional world with humans, animals, stars and other matter. The discovery that our braneworld is embedded in a higher dimensional bulk universe poses an interesting question: does the bulk universe contains a world with lives and other matter which are invisible to us? This invisible world could be the "spirit world" that many religious people believe it must exist.

How did our braneworld come to exist? Cosmological observations have strongly suggested that it started with a Big Bang about 13.7 billion years ago, but its cause is not known. According to physical laws, the only process that can create anything from vacuum is the quantum fluctuation. The Heisenberg's uncertainty principle allows energy to be created spontaneously and then vanish within a short time. The quantum fluctuation has been confirmed experimentally, but only for tiny particles with small amount of energy. Cosmologists assumed that the Big Bang was created by the quantum fluctuation. However, the Big Bang is a gigantic event which requires creation of huge amount of energy. The possibility for quantum fluctuation to create such an enormous energy is nearly zero.

In an article entitled "The String Theory and the Origin of God", an alternative scenario was proposed:

A spirit world was first created from vacuum in the bulk universe. The highly intelligent life (God) in the spirit world then created our braneworld.

The spirit world should obey physical laws, but its matter has entirely different properties than ours. The string theory has been able to explain why they are invisible to us (more info). The physical laws do allow creation of stable matter from vacuum and formation of ordered structures (more info). With the string theory, it is possible to comprehend not only our braneworld, but also the possible spirit world.

What is the origin of the physical laws? We may never have answers to this question. We know our universe has something. Since something cannot come from nothing, we must assume certain things that are self existence (the aseity). In theology, God is the aseity while in science physical laws are the aseity. If we can explain everything in terms of physical laws, we have reached the ultimate goal of science.